STUDY OF GRAIN BOUNDARY CHARACTER

Edited by **Tomasz Tański**
and **Wojciech Borek**

Study of Grain Boundary Character
http://dx.doi.org/10.5772/63043
Edited by Tomasz Tański and Wojciech Borek

Contributors
El Tayeb Bentria, Ibn Khaldoun Lefkaier, Gang Liu, Jinyu Zhang, Jun Sun, Xianglong Yu, Ji Zhou, Fedor Vitalievich Fomin, Nikolai Kashaev, Volker Ventzke, Falk Dorn, Nikita Levichev, Ryan Ganipan Banal, Rolf Sandström, Cheng Yan, Xiaofei Zhu, Guangping Zhang, Vladimir Makin

Published by InTech
Janeza Trdine 9, 51000 Rijeka, Croatia

Publishing Process Manager Dajana Pemac
Technical Editor SPi Global
Cover InTech Design team

First published December, 2016
Printed in Croatia

Additional hard copies can be obtained from orders@intechopen.com

Study of Grain Boundary Character, Edited by Tomasz Tański and Wojciech Borek
p. cm.
Print ISBN 978-953-51-2861-8
Online ISBN 978-953-51-2862-5

For EU product safety concerns:
IN TECH d.o.o., Prolaz Marije Krucifikse Kozulić 3, 51000 Rijeka, Croatia,
info@intechopen.com or visit our website at intechopen.com.

Meet the editors

Prof. Tomasz Tański is the head of the Center for Nanotechnology and vice director of the Institute of Engineering Materials and Biomaterials at the Silesian University of Technology, a member of the Metallurgy Section of the Metallurgy Committee of the Polish Academy of Sciences, and a specialist in nonferrous alloy materials, composite materials and nanostructures, manufacturing engineering, surface, the study of the properties, and in structure engineering materials. He authored or coauthored more than 200 scientific publications worldwide including 12 monographs and books and more than 40 publications in the Philadelphia list, won 18 awards and honors, national and international, is and/or was a supervisor or contractor of more than 15 research and didactic projects in Poland and abroad, and is a reviewer and promoter of numerous scientific papers, including 4 doctoral researches in the field of nanotechnology.

Dr. Wojciech Borek, PhD and MSc in Engineering, is an assistant professor in the Institute of Engineering Materials and Biomaterials at the Silesian University of Technology in Gliwice, Poland, and a member of the association of Polish Cluster of Innovative Forging Technologies "HEFAJSTOS." His scientific interests include materials science, heat treatment, thermomechanical treatment, plastic deformations, and Gleeble simulations. He is an author and coauthor of ca. 50 scientific publications worldwide including 2 chapters in books and more than 15 publications in the Philadelphia list, he won 10 awards and honors, national and international, he is and/or was a contractor of more than 5 research and didactic projects in Poland and abroad, and a reviewer and copromoter of numerous scientific publications, including 2 doctoral researches.

Contents

Preface IX

Chapter 1 **The Effect of Impurities in Nickel Grain Boundary: Density Functional Theory Study 1**
Iben Khaldoun Lefkaier and El Tayeb Bentria

Chapter 2 **Survey of Creep Cavitation in fcc Metals 19**
Rolf Sandström and Junjing He

Chapter 3 **Formation Mechanism and Elimination of Small-Angle Grain Boundary in AlN Grown on (0001) Sapphire Substrate 43**
Ryan G. Banal, Masataka Imura and Yasuo Koide

Chapter 4 **Grain Boundary in Oxide Scale During High-Temperature Metal Processing 59**
Xianglong Yu and Ji Zhou

Chapter 5 **Grain Boundary Segregation in Nanocrystructured Metallic Materials 79**
Jinyu Zhang, Gang Liu and Jun Sun

Chapter 6 **Effect of Microstructure Transformations on Fatigue Properties of Laser Beam Welded Ti-6Al-4V Butt Joints Subjected to Postweld Heat Treatment 111**
Fedor Fomin, Volker Ventzke, Falk Dorn, Nikita Levichev and Nikolai Kashaev

Chapter 7 **Grain Boundary Effects on Microstructural Stability of Nanocrystalline Metallic Materials 143**
Xiaofei Zhu, Guangping Zhang and Cheng Yan

Chapter 8 **Control the Metal Grain Boundary Recrystallization Evolution by the Laser Radiation Electric Field Strength Direction Under Cyclic Thermal Loading 161**
Makin Vladimir Sergeevich

Preface

The grain boundary in a solid material is an interface between two crystals (grains) of the same phase. The mechanical strength and physical properties of industrial materials not only depend on the properties of their components but also are driven by the transition region between two adjacent crystals (boundaries), in particular the chemical constitution and structure of the boundaries. The majority of conventionally used materials are generally polycrystalline; thus, their properties are related not only with the size of grains but also with the grain boundaries. Moreover, modern technological requirements for materials that are "free from defects" and those with high structure defect density make it essential to thoroughly understand the nature of grain boundaries and their influences on the physical and mechanical properties.

Tomasz Tański and Wojciech Borek
Institute of Engineering Materials and Biomaterials
Silesian University of Technology
Gliwice, Poland

Chapter 1

The Effect of Impurities in Nickel Grain Boundary: Density Functional Theory Study

Iben Khaldoun Lefkaier and El Tayeb Bentria

Additional information is available at the end of the chapter

http://dx.doi.org/10.5772/66427

Abstract

By means of density functional theory, we investigate the effect of impurities on structural, electronic and mechanical properties on Nickel Σ5 grain boundary (GB) and its free surface, by studying the effect of 11 transition metal impurities and 8 light elements. The calculation of segregation energy, cohesive energy, formation energy, GB embrittling potency and theoretical tensile strength combined helps us to give accurate conclusions about the effect of these impurities and to compare them with the available experimental and theoretical results. We used the obtained results that are on "equal footing" to establish some correlations and trends. We also confirmed that sulfur and oxygen are the most embrittling elements in Nickel GB in accordance with established literature results and that transition metal elements have a general tendency to segregate to the grain boundaries in a moderate way. Unlike the studied light elements, these elements tend to strengthen the Ni grain boundaries, especially W and Te.

Keywords: nickel grain boundaries, DFT, tensile strength, impurities segregation

1. Introduction

The theoretical investigation of mechanical properties of materials, by means of first-principle calculation and molecular dynamics, has known remarkable advances over the past 15 years [1]. These investigations relate directly to the intergranular behavior, which together with dislocation-based mechanisms determine mechanical response of materials with granular structures. Molecular dynamics calculations can give fundamental understanding of the effects of impurity segregation at GBs in order to predict the mechanical response of materials and their tenability as a function of alloying additions. On the other hand, first-principles calculations can be quite effective in developing this understanding. We can assess whether an impurity is a cohesion enhancer and causes the strengthening of the GB or is an embrittler and induces the weakening of the GB, while experimental techniques are difficult to investigate one impurity alone without the influence of other impurities.

First-principle calculations are widely used for metals and alloys. Grain boundaries of iron make a large part of these investigations. Impurity segregation to the GB in iron is described and understood relatively well, both theoretically and experimentally. In contrast, this is not the case of GB in nickel. The experimental information is quite rare [2]. In recent years, many efforts have been devoted to theoretically determine the effect of impurities on nickel grain boundaries, motivated by the industrial challenges facing fabrication and aging of nickel-base superalloys [3].

It is already well known, both experimentally and theoretically, that sulfur acts as an embrittling element in nickel grain boundary and boron acts as enhancing element [3]. Some other elements like phosphorus are not that clear from first-principle calculations and became a subject of debate between strengthen and embrittling [4]. In contrast to light elements, influence over Ni grain boundaries of alloying elements took less attention [5]. The main elements for this purpose are transition metals, and the most important of them are elements of columns IV–VII and periods 4–6 of the periodic table.

Even though some of these element impurities effects have been studied theoretically, it is remarkable that there are still no comparisons between the effects of theses element in the GB. This is because it is hard to make comparison between results of different works due to the differences in impurity concentrations, GB models, number of atoms per unit cell, methods of calculations and type approximations used. We think that it is important now to have a systematic study to find the most enhancing and the most embrittling elements on Ni GBs. This knowledge can be used also to optimize the material properties as we are trying to understand the mechanisms and effects of different impurity atoms.

This chapter is divided in two parts. Part I is about the methodology of our calculations. Part II describes and discusses the obtained results, and it is organized as follows. First, details of calculation of the effect of Vanadium and Niobium in Nickel GB are exhibited. The effect of light elements "B, P, O, N, Al, Si, S, C" in NiΣ5 GB and surface and also of transition metal elements "Ti, V, Cr, Mn. Zr, Nb, Mo, Hf, Ta, W, Re" in NiΣ5 GB and surface are presented and discussed. Finally, a main conclusion is drawn.

2. Model, methods and computational details

This chapter describes the Nickel Σ5(210) grain boundary models used in these work. Since we are dealing with 2D defect, the constructed structure is not evident as crystals. Beside this, computational parameters used in different models are presented with the precision required and explanation the choice of methods of calculations and the adequate approximations. We briefly present how to calculate free surface energy, grain boundary energy and the tensile strength for a clean and impurity decorated GB.

http://dx.doi.org/10.5772/66427

2.1. Computational details

Total energy calculations and geometry optimizations have been carried out using pseudo-potentials plane-wave method as implemented in Cambridge serial total energy package (CASTEP) [6].

Local density approximation LDA CA-PZ is used for the exchange-correlation potentials (CA-PZ: Ceperley-Alder [7], data as parameterized by Perdew and Zunger [8]). Norm-conserving pseudopotentials (NCP) [6] are used with 720 eV plane-wave cutoff energy for all calculations. The valence electronic configuration of Ni is 3d84s2, and all other transition metal elements are treated including the d shell in valence bonds. The k-points sampling of Ni unit cell was carried out using 8 × 8 × 8 Monkhorst Pack mesh grid, which corresponds to different sampling on different model such as 4 × 4 × 1 k-points sampling for the 22 and 80 atom model and 3 × 5 × 1 for the 44 atom model of our grain boundaries. **Figure 1** presents a sample of 44 atom model used in segregation energies; substitutional segregation sites are sited from 1 to 11. This optimized values lead to an acceptable accuracy for the calculation of the total energy since we are calculating only the difference in total energy between two calculations; thus, a precision of 1 meV is sufficient. In the calculation of density of state and Mullikan population analyses which depend on the derivative of the total energy, we used more strict tolerance such as 6 × 6 × 2 in k-point sampling. The calculations assure a high level convergence of the total energy difference with respect to the number of atoms within 10^{-6} eV for total energy and maximum Hellmann-Feynman force within 0.1 meV/Å for pure Ni Σ5(210) GB and energy difference of 2×10^{-5} eV per atom for segregation and tensile test calculations.

For reasonable and fast convergence of the total energy, electronic occupancies were determined according to a Gaussian scheme with an energy smearing of 0.2 eV. The Pulay scheme of density mixing was used for self-consistent field (SCF) calculations [9].

In this work, we are dealing with Nickel GBs and transition metal impurity mainly, and they provide unique challenges to compute the electronic structure and for some pseudopotentials cause a lot of trouble to converge to an acceptable criteria. The choice of LSDA, norm-conserving pseudopotentials, came after a long study; we have done based mainly of the convergence of the total energy.

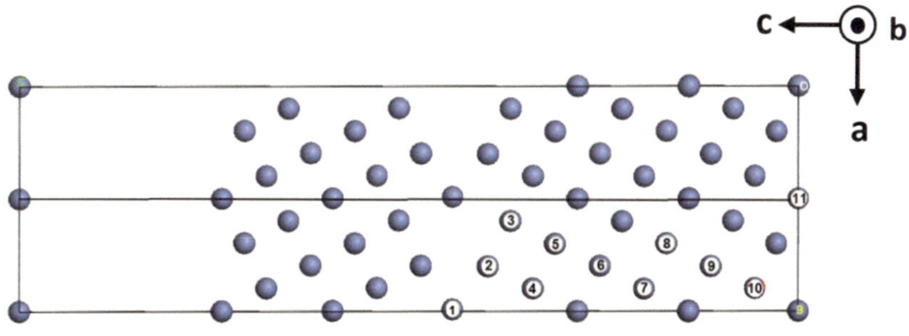

Figure 1. Unit cell model of Ni Σ5(210) symmetrical tilt grain boundary, model used in segregation study. Unit cell shapes are shown by solid lines. Axes directions and orientations are also presented. Atomic sites used in segregation are indicated by numbers (0–11).

Ultrasoft	a (Å) Spin	a (Å) Non-spin	Magnetic moment μ	Norm-conserving	50 SCF step (s)
LDA CA-PZ	3.516	3.451	0.740	LDA	2077.780
GGA-PBE	3.553	3.547	0.660	LSDA	7875.180
GGA-PW91	3.550	3.544	0.640	X-time	3.790
Exp	3.524			Ultra-soft	
NC-PP	Spin	Non-spin	Magnetic moment μ	LDA	2057.780
LDA CA-PZ	3.518	3.506	0.760	LSDA	7851.480
GGA-PBE	3.451	3.440	0.720	PAW-LSDA	2404.180
GGA-PW91	3.430	3.443	0.680	28kpt-600	

Table 1. Nickel lattice parameters in Å with different computational parameters for the magnetic and nonmagnetic case, and the time in seconds for the first 50 SCF cycle for different LDA combinations using 4 cores 3.4 GHz.

Table 1 presents the calculated lattice parameters for Nickel single crystal for different pseudopotentials and methods. As expected, for both approaches "ultrasoft and norm-conserving," LDA underestimates the lattice parameter 3.524 Å and the GGA overestimates it. The error for all methods is acceptable and has a maximum value of 2.66%. Moreover, the non-spin treatment results shrinking of lattice parameter. The method that most predict the lattice parameter is LSDA using norm-conserving pseudopotentials with error of 0.17%.

From **Table 1**, we can also see the CPU time cost for 50 SCF steps using four cores 3.4 GHz and 32 GB RAM. The spin treatment costs about four times that the non-spin calculations, which explains why many calculations in the past were expensive to perform with spin treatment, particularly when the lattice parameters have close values. In order to compare the CPU time used in this work, we perform calculation with the "PAW method" implemented in VASP package, which is the most used method for GBs calculation in Ni. We use the same GB model with the closest parameters. The result shows that PAW is about two to three times much faster than norm-conserving.

We have to mention that the choice of LSDA using norm-conserving was not based mainly on lattice parameters nor on CPU time, but on convergence of total energy for transition metal element, especially during tensile test. The majority of other conducted methods fail to converge when the separation distance between the two surfaces is larger than 3 Å "during tensile test." Norm-conserving with LSDA was the best combination that converges well to the required tolerance in all cases.

2.2. Fundamental grain boundaries parameters

2.2.1. Grain boundary and free surface energies

We introduce the GB energy (γGB) and free surface (FS) energy (γFS) to characterize GB cohesive properties, which are defined as the energies needed to create a GB and FS in the bulk [10]. They are given by the following relations:

$$\gamma^{GB} = \frac{E_{tot}^{GB} - E_{tot}^{Bulk}}{2\,S^{GB}} \text{ and } \gamma^{FS} = \frac{E_{tot}^{FS} - E_{tot}^{Bulk}}{2\,S^{FS}} \tag{1}$$

where E_{tot}^{GB}, E_{tot}^{FS} and E_{tot}^{Bulk} are total energies of the GB, FS and bulk system, respectively, and SGB and SFS refer to GB and FS areas in the GB model. In order to obtain all energies on equal footing and to make suitable comparisons, the calculations of E_{tot}^{Bulk} and E_{tot}^{FS} for the unperturbed Ni FCC ferromagnetic are performed with equal number of atoms per model.

2.2.2. *Segregation and binding energies*

The impurity segregation energy is defined as the difference between the total energy of a system $E_{I\text{-}FS/GB}$ with the impurity in the surface layer (or GB), and the energy E_{Bulk} with the impurity is in the bulk:

$$E_{I/GB} = E_{I-FS/GB} - E_{Bulk}. \tag{2}$$

Therefore, the sign convention is that the negative segregation energy corresponds to impurities that want to segregate [5].

According to the Rice-Wang model [11], the binding energies of the impurities at the grain boundary ΔE_b^{GB} are defined as follows:

$$\Delta E_b^{GB} = E_{I/GB} - E_{GB} - E_I \tag{3}$$

where $E_{I/GB}$ is the total energies of the GB system with segre gated impurity atoms, E_{GB} is the energy of clean GB system, and E_I is the energy for one isolated impurity atom. The larger negative value of the binding energy of the impurity at the GB means stronger bonding between the impurity and nickel atoms.

So in order to calculate these two properties (segregation and binding energies), we need to use six models which are:

- Fully relaxed clean grain boundary, E_{GB}.
- Fully relaxed grain boundary with segregated impurity atoms, $E_{I/GB}$.
- Fully relaxed clean grain boundary with the impurities deep in the bulk region, $E_{I/GB.bulk}$.
- Relaxed clean free surface, E_{FS}.
- Relaxed free surface with segregated impurity atoms, $E_{I/FS}$.
- A monolayer of impurity atoms, E_I, in positions which correspond to the impurity atoms at the grain boundary.

2.2.3. *Theoretical tensile strength*

The theoretical (ideal) strength of a crystal as defined by reference [12] is determined by "*the maximum stress at elastic instability (yield or break) when applying an increasing stress to an infinite perfect crystal. It forms an upper limit to the strength of a real crystal* (**Figure 2**)*, which is of both scientific and engineering value. The theoretical strength is an intrinsic material property, which is*

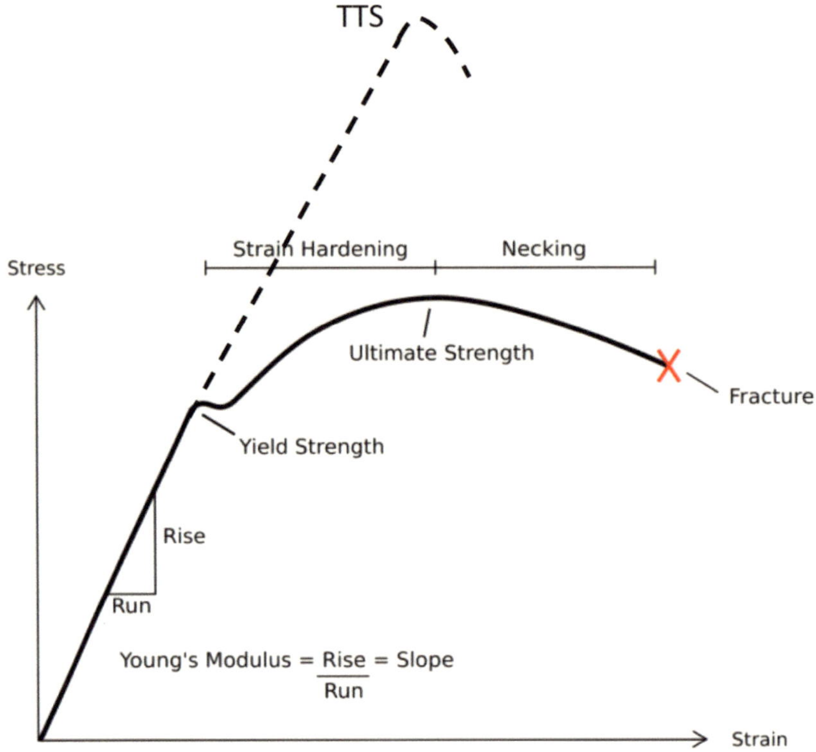

Figure 2. Typical stress vs. strain diagram indicating the various stages of deformation, open source image (modified).

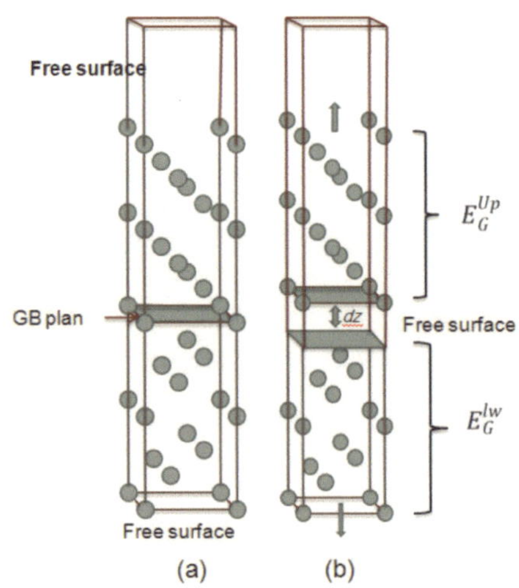

Figure 3. Representation of the GB fracture, the GB is separated in the weakest region, which corresponds to the (210) surface that attach the two grains, dz presents the separation distance.

determined by the behavior of valence electrons and ions. Similarly, the theoretical strength of an ideal defective system containing only one defect such as a point defect, an interface, a grain boundary, etc., can be determined as the maximum stress required to reach elastic instability under increasing load without introducing extrinsic dislocations or cracks" (**Figure 2**) [12].

Here, we recall how to calculate the cohesive energy 2γ and the tensile strength σ_{Max} (equal to the maximum tensile stress), as presented in [13]. We set a fracture plane that gives the minimum cohesive energy; then, the upper and lower half crystal blocks are rigidly separated by five equal increments (**Figure 3**). Each time in the separation process, we perform structure relaxation of the GB region, while fixing atomic layers close to the free surface in order to mimic the bulk structure. Then, the cohesive energy 2γ of the GB is the difference between the two total energies:

$$2\gamma = E_{GB} - E_{S\text{-}GB} \tag{4}$$

EGB is the energy of the GB without separation (point 0), and ES-GB is the total energy for which the separating distance is so large that it does not change any more, typically after 0.5 nm separation.

The maximum tensile stress is calculated as follow. A simple function $f(x)$ is fitted to the calculated total energy versus separation distance x

$$f(x) = 2\gamma - 2\gamma\left(1 + \tfrac{x}{\lambda}\right)\exp\left(-\tfrac{x}{\lambda}\right). \tag{5}$$

Here 2γ and λ are fitting parameters. λ is defined as the Thomas-Fermi screening length. This function is known as the universal binding curve proposed by Rose et al. [14]. It describes well the bonding nature between atoms and constitutes the best fit of binding energies versus atomic distances for the metallic systems. The tensile stress is the derivative of $f(x)$:

$$f(x) = \frac{2\gamma x}{\lambda^2} e^{-\frac{x}{\lambda}}$$

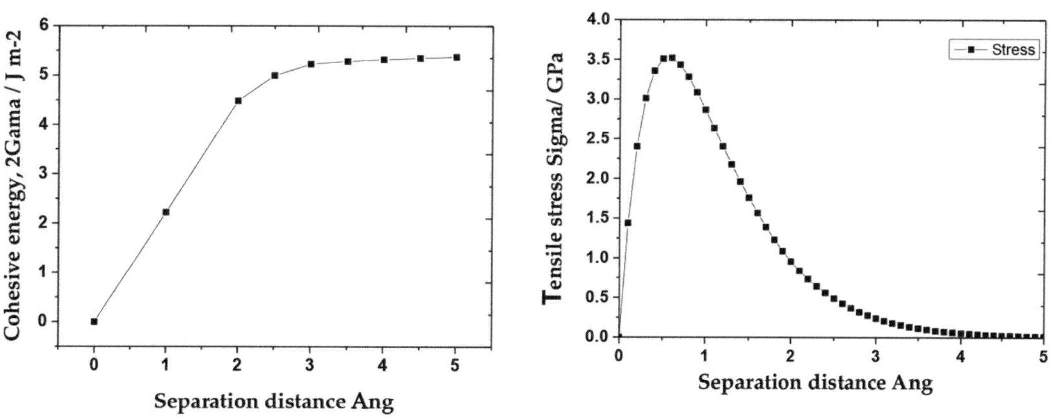

Figure 4. The variation of cohesive energy and tensile stress for FeΣ3(111) GB with boron impurity segregate in site 0.

The maximum of $f'(x)$ is at $x = \lambda$ and corresponds to the maximum theoretical tensile stress or tensile strength σ_{Max}; therefore,

$$\sigma_{Max} = f(\lambda) = e^{-1}\frac{2\gamma}{\lambda} \quad (6)$$

The **Figure 4** represents a calculated example that shows the variation of total energy in Fe Σ3(111) GB with boron impurity segregate in site 0, and the value of the 2γ is 5.41 J m^{-2}; thus, the cohesive energy γ is 2.70 J m^{-2}. The tensile strength is plotted by evaluating the derivative of Rose function $f'(x)$ with function of separation distance x.

3. Results and discussion

3.1. Nickel Σ5 STGB

In order to verify whether the present GGA norm-conserving pseudopotentials (NCP) basis sets are suitable to our Ni GB model, we calculated surface and grain boundary energies presented in Eq. (1) of Nickel GB and compare them with known results of previous experimental and theoretical works (**Table 2**). Indeed, our result agrees well with both experimental results and previous theoretical calculations (see **Table 2**). Thus, we have a strong confidence in the constructed GB configuration of this work.

The calculated total DOS (TDOS) for the grain boundary model (**Figure 5**) appears to have similar characteristics to that of the bulk, that is, the trend of the plot and the density of electrons values are close to each other, and also that the d-bonds are the responsible for bonding, which means that there is no major change in the metallic bonding.

Now we study the difference between Ni atom in perfect bulk and Ni atom in the GB region. From **Figure 5**, we take site 1 as example, because it represents well the different in environments, and we remark that there is a DOS shape compression of the Ni at GB. The difference is much remarkable for the d states that vary from −5 to +3 eV for Ni in the bulk and narrowed for the site1 from −5 to +0.5 eV (**Figure 5**). This compression happened more in the conduction which mean that the GB region reduces electric conduction (barriers) because we know that the electric conduction depends on the electron density around Fermi level [15].

Chen et al. [16] calculate partial DOS (PDOS) for different layers and show that the GB has a significant effect on the shape of d-DOS, and show that at layer 4 the d-DOS became indistinguishable from that calculated for the bulk atom; we could conclude that the effect of the grain boundary in Ni binding disappears totally in the 4th layer.

J.m^{-2}	Our NCP	PP-PAW	US-GGA	Exp. (polycrystal)
GB energy	1.23	1.23[a], 1.43[b], 1.33[c]	1.41[e]	0.93[d], 1.24[f]
FS energy	2.53	2.34[a], 2.65[b], 2.29[c].	2.40[e]	2.59[d], 2.02[f]

a Ref. [18]. b Ref. [20]. c Ref. [13]. d Ref. [5]. e Ref. [29]. f Ref. [30].

Table 2. Calculated grain boundary (GB) energy and free surface (FS) energy of Ni Σ5(210) GB (in J m^{-2}) [28].

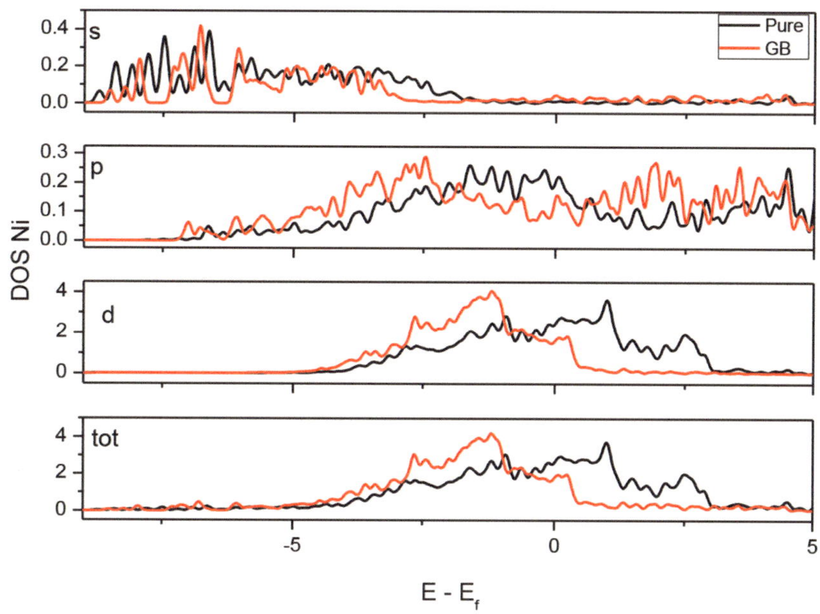

Figure 5. Total and partial density of stated for Ni atom in bulk and in site 1 of the grain boundary.

In both cases, the magnetic moments of Ni increase when we move from GB until it reach the bulk value from the 6th layer on [10], which is also remarked here.

3.2. The effect of light elements "B, P, O, N, Al, Si, S, C" in NiΣ5 GB

In this part, we discuss our results of tensile strength which is uncovered topic for these eight light elements except B and S which is already calculated [17, 18]. We compare our results of segregation and binding energy with the available experimental and theoretical calculations.

3.2.1. Phosphor impurity effect

Controversy regarding the role of P in the Ni GB, Geng et al. [19] proved that P is an embrittler to Ni GB by means of atomic distances, electronic structures and the RWEP. In contrast, Masatake et al. [20] declared that P has a beneficial effect on the Ni GB cohesion, which is in contradiction with previous calculations. While Všianská et al. showed that interstitially segregated P has none or negligible strengthening effect on Ni GB by studying the RWEP. Liu et al. [4] decide to reach a more profound conclusion by calculation of phosphor effect with function of concentration. They found that when the concentration of P is relatively low (0.25–0.5 monolayer, i.e., Np = 1–4; **Figure 6**), P tends to bond strongly with the neighboring Ni atoms, which is beneficial to the GB cohesion.

In the meanwhile, P draws charge from these Ni atoms and removes electrical charges from the Ni–Ni bonds to weaken them. As the P concentration increases (Np = 5–7), P atoms get close and exert a repulsive interaction on each other, thus result in a thin and fragile zone in

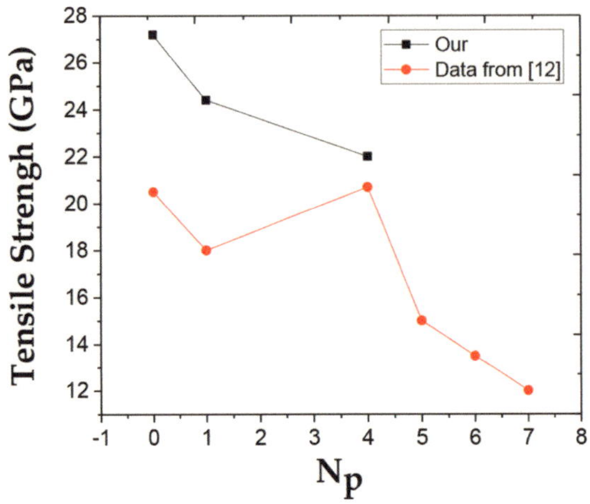

Figure 6. The variation of TTS with function of concentration. The case of NP = 1 represents the GB with 1 P atom in layer 0 or 2, and the case of NP = 4 represents the GB with 4 P atoms in layer 0 or 2. In the range of NP = 5–7, layer 0 is fully occupied by 4 P atoms and layer 2 is occupied by 1–3 P atoms, data from Ref. [4].

the GB. Our calculation corresponds to configuration 2P (Np = 2) corresponding to 0.5 atom/ML at GB see [4], which is favorable from the energetic point of view of Liu (**Figure 6**). The largest segregation energy in Liu is −1.45 eV correspond to 4p concentration. This value is in agreement with ours about −1.77 eV for 1p concentration. The results of tensile strength are different from ours due to the different methods and approximation used for TTS calculation. Moreover, our results for segregation energy are much closer to Všianská et al. (Eseg = −1.6 eV). We found that P has embrittlement effect for this concentration, with positive RWEP, confirmed by the value of TTS (24.4 GPa) and with little decries of cohesive energy to 3.44 eV. For concentration 4P (Np = 4) (1atom/ML), we calculate the TTS, appear to be much lower than Liu value and correspond to 22 GPa with cohesive energy of 3.3 eV that confirm the increase of the embitterment effect with the increasing of the P concentration. Our results for two P concentrations show that there is no enhancing effect of phosphor impurities and acts always as embrittler.

3.2.2. Oxygen and sulfur impurity effect

Oxygen and sulfur have very destructive effect on the Ni GBs. This is clear from tensile strength values presented in, which reduce TTS by 23%. Sulfur is always considered as the most damaging embrittler due to its inevitable existence sometimes in industrial processes. Many experimental and theoretical studies were carried out about its effect [10, 13]. In the other hand, oxygen didn't get that attention at least for theoretical calculation. Here we try to focus more on the effect of oxygen and mention some literature reviews.

Our results show that oxygen impurity present a clear embrittlement to the Ni GB, with the strongest reduction in the cohesive energy that reaches 3.08 J/m^2 of Ni GB. The TTS presents a value of 21.1 GPa and the largest RWEP (1.48 eV/atom). Všianská and Šob [10] explain that by the fact that isolated oxygen atom has a large magnetic energy, which is the reason for making the binding energy of oxygen very large and cause a GB expansion which leads to embrittlement. Furthermore, some experimental observations about the effect of oxygen in NBS by Bricknell and Woodford [21] describe some results and observations on air embrittlement of a commercially pure nickel (Ni200) and show that oxygen was the damaging species and that nitrogen was innocuous. They mention that "*the high concentration of sulfur at the boundaries made oxygen detection by Auger analysis extremely difficult. However, direct evidence for grain boundary oxygen penetration was provided by the formation of various oxygen containing compounds, which were readily observed using scanning electron microscopy*" [21].

Yamaguchi et al. [13] perform comparison between calculated tensile strength for different sulfur impurity concentrations and compare them with experimental ultimate tensile strength with the same concentration of FCC Ni GB [13]. One can see that the order of the tensile strength largely differs between experiment and calculation, and both of strengths are reduced by one order of magnitude with increasing sulfur concentration [13]. The discrepancies in values are due to many reasons. The most plausible of them is the fact that DFT simulations are done for a small model of one symmetrical tilt grain boundary without taking in account the dislocations, while experimental fracture occurs at various kinds of grain boundaries (random grain boundaries, etc.) associated with dislocation emitting. Considering these facts, the agreement between behavior of TTS calculations and experiments seems to be reasonable.

3.2.3. Boron impurity effect

Masatake et al. [20] have calculated the embrittling potency energy of some light elements and show that boron has the best enhancing value for these impurities. This is also the case here. Later then, with 80 atom Ni FCC model, Kart and Cagin [18] investigated the effects of boron segregation at the Ni $\sum 5(210)$ GB in the four interstitial possible positions. They found that increment of the boron atom in the GB site 0 increases the value of the theoretical tensile strength for Ni. Using GGA approximation, they show that the stresses for the GB including one, two, three and four boron atoms are evaluated as 24, 26, 25.8 and 27 GPa, with 23 GPa for the pure GB. Our calculations are relatively in a good agreement, with 37.6 GPa for two boron atoms in GB and 27.6 GPa for the pure Ni GB. The difference in TTS results is due essentially to different TTS calculation methods, which is with GGA approximation known to give lower TTS value than our method approximation (LSDA). All these works agree together that boron is an enhancer to Ni GB. In our calculation, boron presents the highest tensile strength and also the highest RWEP of −0.94 eV. This makes him the best impurity type in the considered eight light elements.

3.2.4. Aluminum and silicon impurity effect

These two elements are juxtaposed on the same colon, that is, they have the same electron number in valence band. These elements take the less attention in the computational side due to their indecisive effect and controllability of their existence in experiments [22].

Geng et al. [19] have expect that aluminum and silicon, which have similar atomic size and bonding as phosphorus, would be embrittler for Nickel $\sum 5(210)$ GB. Later on, Všianská and

Šob [10] have calculated the segregation energy and embrittlement potency and show that, contrary to the Geng prediction, Si acts as enhancer with Rice WEP −0.41, and Al has no significant effect in Ni GB with RWEP −0.03.

For Silicon, our results confirm Šob findings, but they go more to Geng prediction for Al which presents more embrittlement character with RWEP +0.16. Beside Geng and Šob approaches, our results expected to be more accurate with the calculation of the cohesive energy and the TTS, which show embrittlement of Al impurities to 24.6 GPa, and a little enhancement of Si impurity to 28.2 GPa, and validate the RWEP values.

3.2.5. Carbon and nitrogen impurity effect

Siegel and Hamilton [31] have studied carbon segregation and diffusion within a Nickel ∑3 grain boundary and show that interstitial site is preferred for the magnetic and nonmagnetic cases. But the author did not calculate the strengthen effect. In 2008, Sanyal et al. [23] have talked about the strengthen effect of carbon impurity in Nickel ∑5(210) GB using the cohesive energy values. They founds decries of the cohesive energy from 3.60 to 3.54 eV, in which they concluded that C acts as embrittler. Using a 20 atoms model, Masatake et al. [20] have found negative RWEP which means that C acts as cohesive enhancer. Contrarily, Young et al. [24] find positive RWEP with 40 atom models which confirm that C is an embrittler impurity. This result is in agreement with our 2 × 2 64 atoms model, in which we found enhancement in cohesive energy, TTS and negative RWEP. This controversy in results is mainly due to the model number of atoms, and thus, we confirm the remark of [25] that low segregation energy can give wrong values.

For the Nitrogen case, we found that N impurity atom results a positive RWEP and a lower cohesive energy, but with a small enhancement in the TTS. This value of TTS is not in agreement with the RWEP and cohesive energy. RWEP results in [20] agree well with our RWEP finding for the non-spin calculation, taking into account the loose in cohesive energy of a GB with N impurity. Taking in to account the result of Masatake and as mention above that the total energy and forces of some elements mainly for N do not converge to the required criteria for some fracture cases, we can conclude based only on the two factors (RWEP en Ecoh) that N is embrittler.

3.2.6. General tendency

Young et al. [24] have put the order of the impurities effect from most embrittling to most strengthening, the impurity elements are ranked as He, Li, S, H, C, P, Fe, Mn, Nb, Cr, and B. Helium is strongly embrittling (+1.07 J/m^2 lowering of the RWEP), while phosphorus has little effect on the grain boundary (−0.05 J/m^2), and boron offers appreciable strengthening (−0.54 J/m^2 increase in RWEP). These findings are consistent with experimental observations (e.g., He, S and H are known embrittling agents and boron is a known strengthener in nickel-base alloys). This classification is in excellent agreement with our tensile strength results even the paper was conducted on ∑5 twist Ni GB along, which give us more confidence that the study of the effect of impurity in one type of GB could lead us to general result about its behavior in other type of Ni GBs.

3.3. The effect of transition metal elements "Ti, V, Cr, Mn. Zr, Nb, Mo, Hf, Ta, W, Re" in NiΣ5 GB

Light elements are, in general, known as embrittling agents for Nickel, whereas transition metal elements are known as the main enhancing alloying impurities. Even though these elements have been subject of intensive thermodynamic based studies for their effects on Nickel as impurities, there is a lot of discrepancy in results of different models [19]. Nevertheless, there are few works which treat it by quantum mechanical approaches [23, 24]. Razumovskiy et al. [5] calculated the segregation energy, partial cohesive energy and RWEP for W, Zr, Hf, Bi, S, B, Ta and Re. While writing this work, another work from Ref. [3] conducts a study of purpose to the design of Ni-base polycrystalline superalloys by studying the influence of a wider range of transition metal elements on grain boundary segregation and bulk cohesion in Ni ∑5(210) GB.

In this paragraph, we compare and discuss the results of segregation energy and tensile strength for some elements mentioned in other works. Hf impurity shows an important RWEP of −0.61 in agreement with Razumovskiy et al. value of −0.69 eV/atom. Sanyal et al. [23] found that Hf was favored at the GB by +0.8 eV/atom relative to that in the bulk. Hf is experimentally found to strengthen GBs [26]. In this work, we add more information about its effect by TTS calculation. We predict that Hf is not a good enhancer if available alone in the GB. Its existence in Ni GB with other embrittling elements such as sulfur further should be studied. The high segregation energy of Hf may compete S segregation to Ni GB and overcome its undesired effect.

Ta has the highest RWEP among transition metals with a large cohesive energy of +5.22 eV and second largest TTS after W. Ta is hence one of the best enhancer. We found a negative and large value of segregation energy at GB for Ta −0.91 eV. Similar conclusion was also drawn by [5].

If we referee to our calculated results about Cr additions, this element does not prefer to be at the GB region, but rather to the surface. Even though, it presents a positive RWEP. Its effect in the GB if existed is not harmful and doesn't affect considerably the cohesive energy and TTS calculations. The later result is confirmed by the small negative RWEP value of Cr in Young work [24].

W has the largest cohesive energy, the highest TTS and a very important RWEP. Based on these factors, we can say that W is the best enhancer among the 12 considered transition metals elements with small difference with Ta. Represents high cohesive energy, but it does not have the highest tensile strength, in accordance with Tahir's work [27] that cohesive energy does not give full presentation of enhancement/embrittlement.

Based on other first-principles method calculations (full-potential linearized augmented plane-wave method FLAPW). Geng et al. [19] calculated that the cohesive energy for all transition metals elements in order to, unambiguously, predicts the effect of a substitution alloying addition on grain boundary cohesion of metallic alloys. **Figure 7** presents a comparison between our calculated values for the cohesive energy and those obtained by Gang et al. [19]. We notice that even there is a similar trend in both results (e.g., W has the largest cohesive

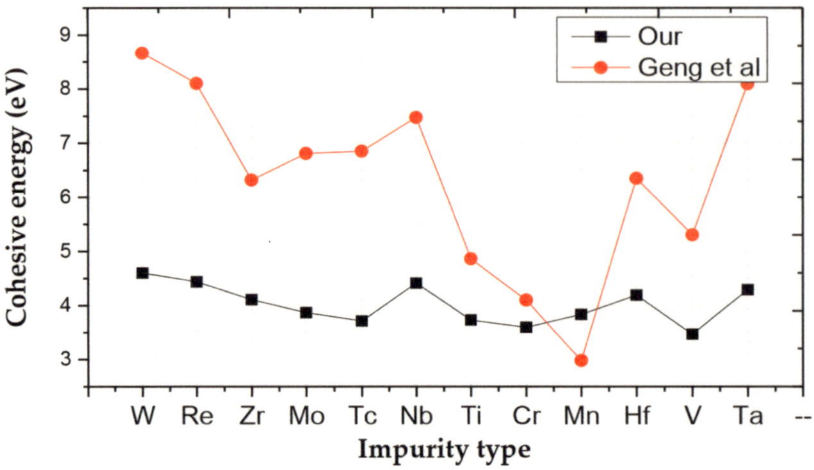

Figure 7. The cohesive energy of Nickel Σ5 GB with impurity type in it, element in X axes represents the impurity type (one atom) generally located on Site1. Data 1 taken from Ref. [19].

energy followed by Re and Ta approximately with the same value and so on), and the cohesive energy values are quite different. **Figure 7** shows that the cohesive energy from our calculations varies in the range [+5.6 to +4.6 eV], while Geng results fluctuate largely from +8.66 to +2.98 eV. Geng results seem to be less accurate than ours. For example, Geng calculated cohesive energy for W impurity at Ni GB and found 8.66 eV, 2.4 time larger than our value and other known pure GB cohesive energies [19]. This value (8.66 eV) could lead to very high value of tensile strength equal to 65 GPa and vice versa 22 GPa for Mn (lower than sulfur), which are simply not reasonable for the magnitude range of Ni-TTS. We think that these fluctuations are first due to the number of atoms (22 atoms) in Geng model, which leads to highest segregation energies and therefore higher effect on the GB TTS. Also, the relaxation of atomic positions was only done in the normal direction to the GB plane and ignored in both lateral and the mirror symmetries in the normal direction to the GB plane (210).

4. Conclusion

In this work, we conduct analyses about the influence of segregated impurities on the properties of nickel grain boundaries. The problematics and controversy of results posed here have a dual character: fundamental and industrial. The selected impurities are eight light elements and eleven transition metals elements of the periodic table. The adopted methodology for this study was based on density functional theory widely used in recent years to predict the mechanical response of materials and their tenability as a function of alloying additions.

After optimization study, we have shown that the norm-conserving approach together with local density approximation is best suited to solve convergence problems as well as to give accurate results in the case of metallic systems. Furthermore, different GB models have been used in order to fulfill the required calculated property.

Calculations permitted us to confirm that light elements (column 13–16, periods: 2 and 3) have strong tendency to segregate at Ni grain boundary due to their small size and high electronegativity. However, their influence on the cohesion of the grain tensile strength (TTS) differs from one element to another. We also confirmed that sulfur and oxygen are the most embrittling elements in Nickel GB in accordance with established literature results.

Our calculations show that the transition metal elements (column 22–25 periods: 4–6) have a general tendency to segregate to the grain boundaries in a moderate way. Unlike light elements, these elements tend to strengthen the Ni grain boundaries, with the exception of Mn which acts as embrittler, in which we studied further in order to explain this case. We have shown that the magnetism of Mn plays an important role in the GB decohesion. Among these transition elements, we have shown that W and Re are the most consolidating for Ni GB and therefore are candidates for counter embrittlement of sulfur.

Author details

Ibne Khaldoun Lefkaier[1] and El Tayeb Bentria[2*]

*Address all correspondence to: ik.lefkaier@lagh-univ.dz

1 Materials Physics Laboratory, University of Laghouat, Laghouat, Algeria

2 Qatar Environment and Energy Research Institute (QEERI), Hamad Bin Khalifa University, Doha, Qatar

References

[1] Erich, W., N. Reza, A. Y. George, Jr., D. B. Jake, M. A. Thomas, V. James, J. C. James, N. Hiroaki, B. S. Judy, F. Clive, C. Mikael, W. Walter and S. Paul. Ab initio calculations for industrial materials engineering: successes and challenges. Journal of Physics: Condensed Matter. 2010;**22(38)**:(384215).

[2] Všianská, M. Electronic structure and properties of grain boundaries in nickel. Diss. Masarykova univerzita, Přírodovědecká fakulta. 2013.

[3] Razumovskiy, V. I., A. Y. Lozovoi and I. M. Razumovskii. First-principles-aided design of a new Ni-base superalloy: Influence of transition metal alloying elements on grain boundary and bulk cohesion. Materials Science and Engineering A-structural Materials Properties Microstructure and Processing. 2015;**107**:(23–40).

[4] Liu W., C. Ren, H. Han, J. Tan, Y. Zou, X. Zhou, et al. First-principles study of the effect of phosphorus on nickel grain boundary. Journal of Applied Physics. 2014;**115**:(043706).

[5] Razumovskiy, V. I., A. Y. Lozovoi, I. M. Razumovskii and A. V. Ruban. Analysis of the alloying system in Ni-base superalloys based on ab initio study of impurity segregation to Ni grain boundary. Advanced Materials Research. 2011;**278:(2011)**:(192–197).

[6] Clark Stewart, J., D. Segall Matthew, J. Pickard Chris, J. Hasnip Phil, I. J. Probert Matt, K. Refson and C. Payne Mike. First principles methods using CASTEP. Zeitschrift für Kristallographie. 2005;**220**:(567).

[7] Ceperley, D. M. and B. J. Alder. Ground state of the electron gas by a stochastic method. Physical Review Letters. 1980;**45(7)**:(566–569).

[8] Perdew, J. P., A. Zunger. Self-interaction correction to density-functional approximations for many-electron systems. Physical Review B. 1981;**23**:(5048–79).

[9] Pulay, P. Ab initio calculation of force constants and equilibrium geometries in polyatomic molecules: I. Theory. Molecular Physics. 1969;**17(2)**:(197–204).

[10] Všianská, M. and M. Šob. The effect of segregated sp-impurities on grain-boundary and surface structure, magnetism and embrittlement in nickel 817-840. Progress in Materials Science. 2011;**56(6)**:(817–840).

[11] Rice, J. R. and J. S. Wang. Embrittlement of interfaces by solute segregation. Materials Science and Engineering: A. 1989;**107**:(23–40).

[12] Lu, G. -H., et al. Theoretical tensile strength of an Al grain boundary. Physical Review B. 2004;**69(13)**:(134106).

[13] Yamaguchi, M., M. Shiga et al. Grain boundary decohesion by sulfur segregation in ferromagnetic iron and nickel-a first-principles study. Materials Transactions. 2006;**47(11)**:(2682–2689).

[14] Rose, J. H., J. Ferrante and J. R. Smith. Universal binding energy curves for metals and bimetallic interfaces. Physical Review Letters. 1981;**47(9)**:(675–678).

[15] Mott, N. F. The electrical conductivity of transition metals. Proceedings of the Royal Society of London. Series A, Mathematical and Physical Sciences. 1936;**153(880)**:(699–717).

[16] Chen, L., et al. First-principle investigation of bismuth segregation at Σ5 (210) grain-boundaries in nickel. Transactions of Nonferrous Metals Society of China. 2006;16(**Supplement 2(0))**:(s813–s819).

[17] Yamaguchi, M., M. Shiga and H. Kaburaki. Grain boundary decohesion by impurity segregation in a nickel-sulfur system. Science. 2005;**307(5708)**:(393–397).

[18] Kart, H. H., T. Cagin. The effects of boron impurity atoms on nickel ∑ 5 (210) grain boundary by first principles calculations. JAMME. 2008;**30:(2008)**:(177–181).

[19] Geng, W. T., A. J. Freeman, R. Wu, C. B. Geller and J. E. Raynolds. Embrittling and strengthening effects of hydrogen, boron, and phosphorus. Physical Review B. 1999;**60(10)**:(7149–7155).

[20] Masatake, Y., S. Motoyuki and K. Hideo. Energetics of segregation and embrittling potency for non-transition elements in the Ni Σ5 (012) symmetrical tilt grain boundary: a first-principles study. Journal of Physics: Condensed Matter. 2004;**16(23)**:(3933).

[21] Bricknell, R. H., and D. A. Woodford. The embrittlement of nickel following high temperature air exposure. Metallurgical Transactions A. 1981;**12(3)**:(425–433).

[22] Sims, C. T. A history of superalloy metallurgy for superalloy metallurgists. Superalloys. 1984;**1984**:(399–419).

[23] Sanyal, S., U. V. Waghmare, P. R. Subramanian and M. F. X. Gigliotti. Effect of dopants on grain boundary decohesion of Ni: a first-principles study. Applied Physics Letters. 2008;**93(22)**:(223113).

[24] Young, G., R. Najafabadi et al. Applications of Ab Initio Modeling to Materials Science: Grain Boundary Cohesion and Solid State Diffusion. 2004;No. LM-04K037. Lockheed Martin Corporation, NY, US.

[25] Lejček, P., et al. Why calculated energies of grain boundary segregation are unreliable when segregant solubility is low. Scripta Materialia. 2013;**68(8)**:(547–550).

[26] Jena, A. K. and M. C. Chaturvedi. The role of alloying elements in the design of nickel-base superalloys. Journal of Materials Science. 1984;**19(10)**:(3121–3139).

[27] Tahir, A. M. Development and validation of a scale-bridging method for simulation of intergranular fracture in body centered cubic metals. ICAMS 2014 (thesis).

[28] Bentria, E. T., I. K. Lefkaier, and B. Bentria. The effect of vanadium impurity on Nickel Σ5(012) grain boundary. Materials Science and Engineering A. 2013;577:(197–201).

[29] Tyson, W. and W. Miller. Surface free energies of solid metals: estimation from liquid surface tension measurements. Surface Science. 1997;**62(1)**:(267–276)

[30] Zhang, L., X. Shu, S. Jin, Y. Zhang and G. -H. Lu. First-principles study of He effects in a bcc Fe grain boundary: site preference, segregation and theoretical tensile strength. Journal of Physics: Condensed Matter. 2010;**22(37)**:(375401)

[31] Donald J. Siegel, J.C. Hamilton. Computational study of carbon segregation and diffusion within a nickel grain boundary. Acta Materialia. 2005;**53**:(87–96).

Chapter 2

Survey of Creep Cavitation in fcc Metals

Rolf Sandström and Junjing He

Additional information is available at the end of the chapter

http://dx.doi.org/10.5772/66592

Abstract

Cavitation plays an important role in plants operating at high temperatures since the cavitation controls the creep failure of engineering alloys. In the past it has been difficult to predict the cavitation behaviour with the help of basic models, since critical models have been missing. Recently new models have been formulated for grain boundary sliding, cavity nucleation and cavity growth to fill this gap. These models are reviewed in this chapter. It is shown that the new models can quantitatively predict cavitation for austenitic stainless steels, where detailed experimental information is available.

Keywords: creep, cavitation, grain boundary sliding, austenitic stainless steels, copper

1. Introduction

At temperatures above 0.4 of the absolute melting temperature T_m, materials are exposed to a slow plastic deformation called creep. For steel 0.4 T_m is about 500°C. The slow deformation takes place also at constant stress. Many important technical units operate at such high temperatures. This applies, for example, to fossil fired power plants that produce most of the world's electric power. Another example is gas turbines in aircrafts where the maximum metal temperature exceeds 900°C.

Due to creep the total strain in the material gradually increases. At the same time the microstructure can also change. For example the strength of many materials used at high temperatures is based on the presence of fine particles that slow down the deformation. There is a thermodynamic driving force for coarsening of the particles, since this will reduce the total surface area of the particles and thereby the surface energy. The deformation of the material and changes in the microstructure will decrease the strength of the material and this is referred to as the formation of creep damage. When the creep damage has reached a certain level, failure takes place.

Polycrystalline materials consist of regions with a specific lattice orientation called grains and the boundaries between them grain boundaries. During the creep deformation, small voids called creep cavities are formed at the grain boundaries. The size for the cavities is of the order 1 μm. The creep cavities are continuously nucleated so their number increases with time. Each individual cavity also grows so their radii increase. The driving force is the same as for coarsening of particles, that is, a reduction of the surface energy. In this way, there is a gradually increasing fraction of the grain boundaries that is cavitated. When the area fraction of cavities has reached a critical value, the cavities join and eventually form cracks that make the material fail.

The fact that creep rupture is mainly controlled by the development of grain boundary cavities has created a large technical interest in cavitation. This interest increased even more in the 1980s, when it was recognised that the appearance of the cavitation could be used to estimate the residual life time of fossil fired power plants. The service time of many power plants were approaching the design life and operators were asking whether it would be safe to continue running the plants. Neubauer found that by observation of the cavitation with the help of the replica technique, the residual lifetime could be estimated [1, 2]. The cavitation was subdivided into four classes (and one for undamaged material): individual cavities to a small extent, individual cavities to a large extent, stringers of cavities and finally microcracks. Replicas were taken at welds, pipe bends and other critical positions. The basic idea was that if damage of one class was detected, the damage did at most correspond to the next class at the next inspection. The method was very successful when it was applied to low alloy steels such as 0.5Cr0.5Mo0.25V and 2.25Cr1Mo, which represented the body of materials of the plants at the time. Unfortunately the method is less applicable to today's materials such as 9 and 12Cr steels, because cavitation appears only at a late stage of life and does not provide the necessary early warning.

The successful technical use of observations of cavitation stimulated a lot of scientific work. This was dominated by empirical approaches to describe the development of the creep damage. The first and perhaps most well-known approach was set up by Kachanow and Rabotnov [3–5]. They simply assumed that the cavities represented voids that reduced the loading capacity. The most interesting feature of the method is that it is consistent with behaviour of creep strain during the tertiary creep, that is, the final stage before rupture [6, 7]. This dependence is nowadays referred to as the omega method [8].

We will now concentrate on basic models for cavitation. Traditionally cavity nucleation has been modelled either as a process of rupturing atomic bonds or of atomic vacancy condensation. For the former approach the estimated threshold stress is orders of magnitude higher than the applied stress, which makes it physically unrealistic because high stresses will be reduced quickly in a creeping material [9]. The condensation of vacancies can be treated with the help of the classical nucleation theory [10]. It is shown that cavity nucleation would be a very rare event at low stresses but becomes frequent above a certain threshold stress [11]. High stress concentration can be formed at grain boundary ledges, grain boundary triple points and particles. Cavity nucleation at particles can be a result of decohesion of particles from the matrix. In practically all models that have been presented, a high stress concentration is needed.

A threshold stress and an incubation time are essential to form a cavity [10, 12]. Contrary to these suggestions nucleation frequently takes place at low stresses and is controlled by strain rather than stress. In agreement with statements in the literature it can be concluded that theories of cavity nucleation have not been fully successful in earlier work [13, 14]. In both these papers excellent reviews are given.

For modelling cavity growth the situation was different but not entirely unproblematic. A diffusion controlled cavity growth was formulated by Hull and Rimmer [15]. The model was later improved in particular by Beere and Speight [16]. It was soon realised that the model gave much larger growth rates than observed experimentally in many cases. Dyson proposed that the cavities should not be able to grow faster than the creep deformation of the surrounding material [17]. This is referred to as constrained cavity growth. Rice developed an explicit model for constrained growth [18]. The idea of constrained growth is now fully accepted. However, the model still tends to overestimate the growth rates. It is now believed to be due to the assumptions in Rice's derivation. This will be analysed in the present paper.

It is evident from the summary above that the modelling of formation and growth of cavities has met considerable difficulties in the past. In recent years important new developments have taken place. With the help of these developments fundamental quantitative modelling of both nucleation and growth of creep cavities has now been possible to set up. In addition some of the difficulties that have been encountered in the past can be understood. It is the purpose of this paper to review these new developments.

2. Grain boundary sliding

Grain boundary sliding (GBS) occurs when neighbouring grains move with respect to each other in shear. The mechanism is illustrated in **Figure 1**. During the creep deformation the two grains have moved 0.8 μm with respect to each other. To observe GBS, the specimen surface has to be scratched, for example, with a knife. When the scratches cross a sliding grain boundary, the two parts of the scratches on the different sides of the grain boundary are displaced. This displacement is a direct measure of GBS.

It is generally accepted that a prerequisite for cavity nucleation is grain boundary sliding. Experiments on copper bicrystals have shown that artificially introduced GBS can dramatically increase the amount of cavitation. Chen and Machlin [20] and Intrater and Machlin [21] exposed bicrystals of copper to either tensile loading or to a combination of tensile loading and shear. The latter alternative gave much larger number of cavities. It is natural that GBS gives rice to cavitation, since any obstacle at the grain boundary such as a particle will give rise to large stress concentrations. In low alloy steels cavities have frequently been observed around manganese sulphides [22]. Since the interface between the sulphides and the matrix is weak, voids are easily formed there. Some papers also suggest that cavities can be formed at carbides, see, for example, Ref. [23].

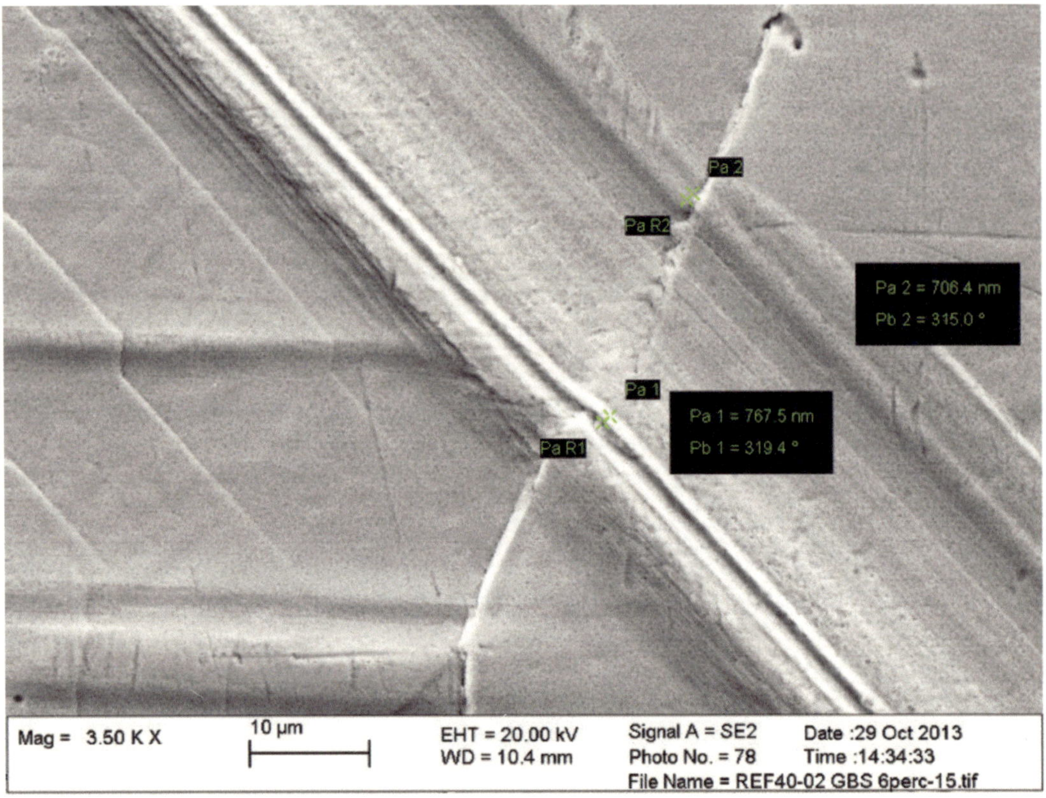

Figure 1. Illustration of grain boundary sliding (GBS) for a copper specimen that has been exposed to 3.3% creep strain during 307 h at 125°C [19]. The grain boundary lies in the southwest-northeast direction. It is crossed by a major scratch, which makes it possible to measure GBS. The grain to the left has moved downwards by 0.8 µm relative to the grain at the right and that is the amount of GBS.

The most convincing argument concerning the central role of GBS in cavity nucleation comes from the creep strain dependence of both GBS and cavity nucleation. It has been observed many times that the displacement u_{GBS} due to GBS is approximately proportional to the creep strain ε, see, for example, Ref. [24]. The first ones to observe this relation were McLean and Farmer [25].

$$u_{\mathrm{GBS}} = C_s(\varepsilon)\varepsilon. \tag{1}$$

$C_s(\varepsilon)$ is a constant that is dependent on the creep strain ε. At the same time the nucleation rate of cavities $\frac{dn}{dt}$ is also proportional to the creep strain rate $\dot{\varepsilon}$.

$$\frac{dn}{dt} = B\dot{\varepsilon}. \tag{2}$$

B is constant. This means that the number of cavities is proportional to the creep strain in the same way as the GBS displacement in Eq. (1). Eq. (2) was first observed by Needham and

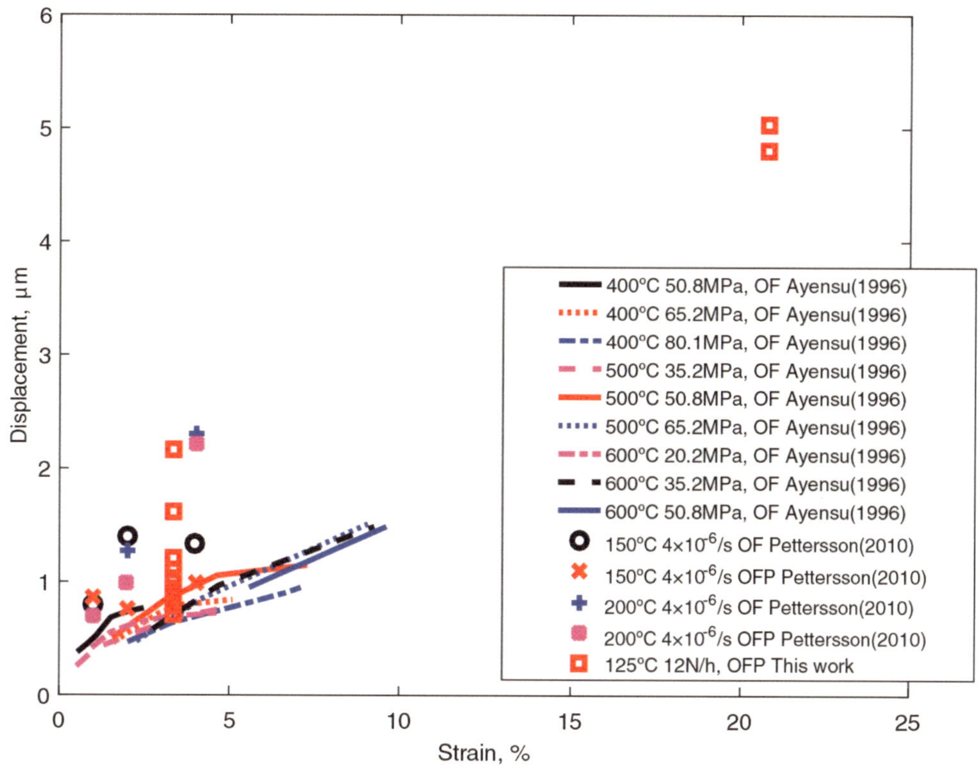

Figure 2. Observed displacements at grain boundaries in copper as a function of strain [19]. Data from Refs. [29, 30] are also shown.

coworkers [26, 27]. For a review, see Ref. [28]. Eqs. (1) and (2) will be derived below. Eq. (2) would be very difficult to explain unless it assumed that the nucleation is controlled by GBS.

Experiments that give the strain dependence of u_{GBS} are illustrated in **Figure 2**.

The displacement u_{GBS} increases as of function of strain in agreement with Eq. (1). C_s is the slope of the curves. Three types of tests are represented in the figure: tests at constant stress, at constant stress rate and at constant strain rate. In spite of the fact that a range of temperatures, strain rates and test methods is covered, the C_s values do not vary very much.

To investigate the influence of GBS on the total strain, Crossman and Ashby [31] developed a finite element model (FEM) for shear stresses. If a free grain boundary is considered, they found that the sliding rates are very high for typical creep stresses and that the grain boundaries could be considered as flaws in the material with respect to GBS. Later Ghahremani [32] transferred the model to tensile stresses, which are typically used in creep testing. In both Refs. [31, 32] a Norton equation for the creep strain rate was considered

$$\dot{\varepsilon} = \dot{\varepsilon}_0 \left(\frac{\sigma}{\sigma_0}\right)^n. \tag{3}$$

σ is the applied stress and n is the creep exponent. $\dot{\varepsilon}_0$ and σ_0 are constants. The percentage creep rate due to grain boundary sliding ϕ was determined

$$\phi = \frac{\dot{u}_{\text{GBS}}}{\dot{u}_{\text{all}}}. \tag{4}$$

$\dot{u}_{\text{all}}$ is the total displacement rate. ϕ was found to take values from 0.15 ($n = 1$) to 0.33 ($n = \infty$) in Ref. [32]. $\dot{u}_{\text{all}}$ can be expressed in terms of the creep rate $\dot{\varepsilon}$

$$\dot{u}_{\text{all}} = \frac{3\, d_{\text{lin}}\, \dot{\varepsilon}}{2\xi}, \tag{5}$$

where d_{lin} is the linear intercept grain size and $\xi = 1.36$ is a pure geometrical factor that explains how the hexagonal grains studied in Refs. [31, 32] should be related to the measured grain size. The factor 3/2 depends on the definition of $\dot{u}_{\text{all}}$. By combining Eqs. (1), (4) and (5) we find the values of the GBS parameter C_s in Eq. (1)

$$C_s = \dot{u}_{\text{GBS}} \Big/ \dot{\varepsilon} = \frac{3\phi}{2\xi} d_{\text{lin}}. \tag{6}$$

Eq. (6) is referred to the *shear sliding model*. Eq. (6) is compared with experimental results for copper in **Figure 3**. The C_s values according to Eq. (6) for the individual tests in **Figure 2** have

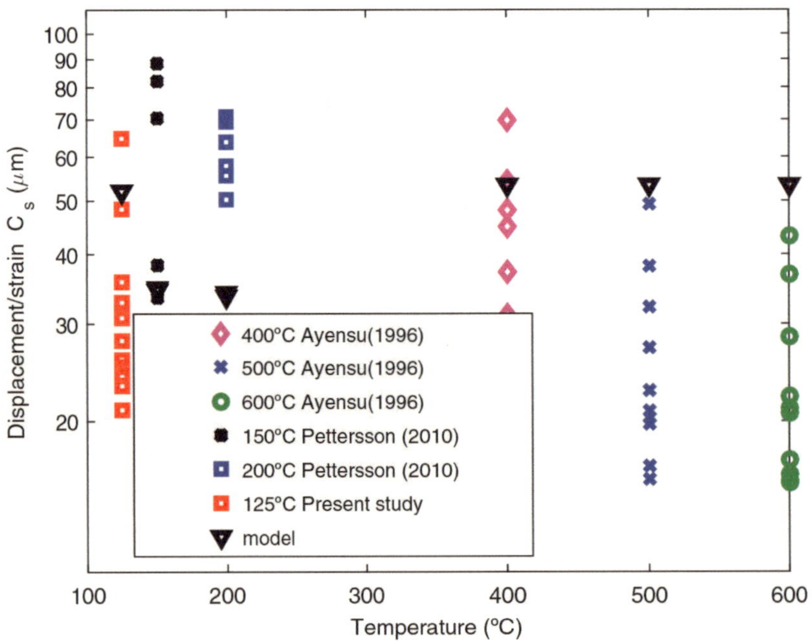

Figure 3. Comparison of modelled Eq. (6) and observed displacements at grain boundaries in copper divided by the creep strain, cf. Eq. (1) [19]. Data from [29, 30] are also shown.

been evaluated in [19]. The model values are about $C_s \approx 50$ μm. These values are slightly high for the creep tests [30], but in range for slow strain tests [29] and constant stress rate tests [19].

For materials with particles in the grain boundaries Riedel has derived a model corresponding to Eq. (1) [33]. The sliding boundary was represented by a shear crack surrounded by creep deforming grains. The model is referred to as the *shear crack model*. Although the author was not very happy with the model, it turns out that it does not give very different results for austenitic stainless steels in comparison with the shear sliding model.

The two models (shear sliding and shear crack models) are compared with the experimental GBS displacements for different austenitic stainless steels [34–38] in **Figure 4**. The shear crack model is compared with the average of all the experimental data, Ave. 1. The shear sliding model does not work very well for large grain sizes so data for such grain sizes [38] are not included in the comparison, Ave. 2. It can be seen from **Figure 4** that C_s values of the correct order are predicted.

From **Figures 3** and **4** it is evident that the shear sliding and shear crack models can describe the experimental data for fcc metals with reasonable precision.

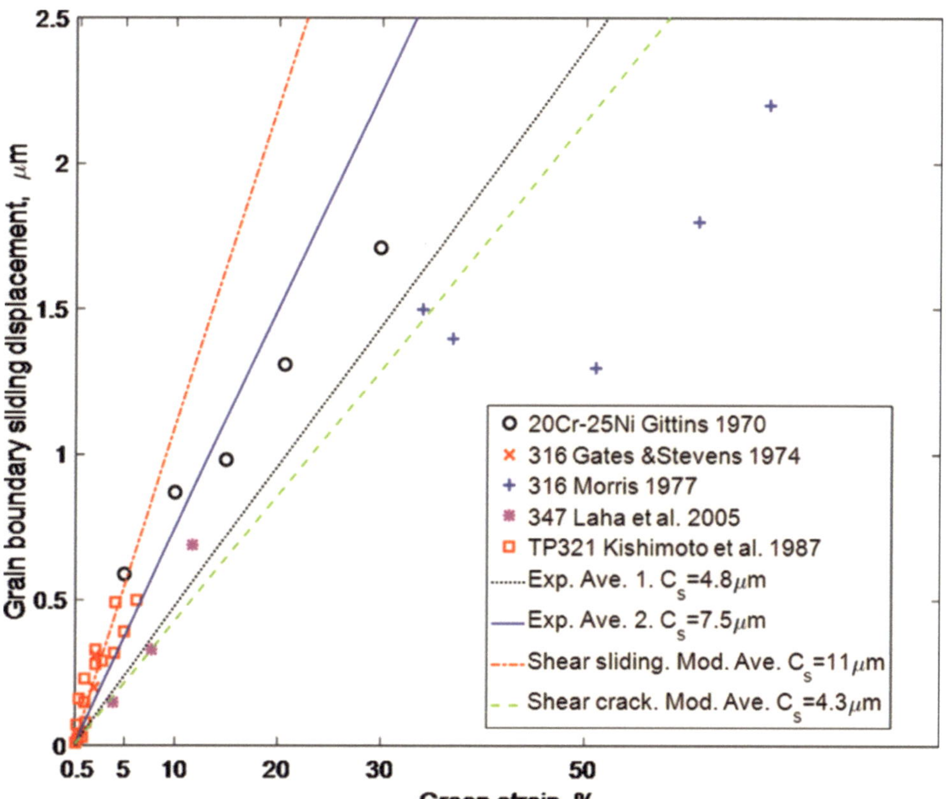

Figure 4. Modelling and experimental GBS displacement as a function of creep strain for different types of austenitic stainless steels, from Ref. [24]. Experimental data Ref. [34–38].

3. Cavity nucleation

3.1. Thermodynamic considerations

The mechanisms for cavitation nucleation have been a puzzle for a long time as explained in the introduction. However part of the explanation came from studies on copper. Pure copper can show extensive cavitation during creep [39], but the number of particles present is so low that they cannot explain the large number of cavities. Lim suggested that it was the substructure of the dislocations that could nucleate the cavities [40]. He also presented a model that can be used to demonstrate whether a nucleation mechanism is thermodynamically feasible or not. He assumed that pile ups of grain boundary dislocations generate the necessary high stresses for the nucleation. Since these high stresses are stationary as a result of the creep process, it avoids the problem of fast stress relaxation in many models. Lim's model is fairly complex and details in the model will not be given here. When a cavity is formed the free energy ΔG is changed in a number of ways that are represented by the terms in the following equation [19]

$$\Delta G = -r^3 F_v \sigma_{appl} + r^2 F_s \gamma_s - r^2 F_{GB} \gamma_{GB} - (\Delta G_1 + \Delta G_2 + \Delta G_3). \quad (7)$$

γ_s and γ_{GB} are the surface and grain boundary energies per unit area. $F_v = 2\pi/3\ (2–3\cos\alpha + \cos^3\alpha)$, $F_s = 4\pi(1–\cos\alpha)$, $F_b = \pi\sin^2\alpha$ and $F_v' = 1.5\ F_v$, where α is half the tip angle of the cavity. The first term in Eq. (7) is the work done by the applied stress. The second and third terms represent the modification in the surface and grain boundary energies. The fourth term is the decrease in the strain energy. ΔG_1 is the change in the line energy of the grain boundary dislocations (GBD). ΔG_2 is the interaction energy between the remaining and the consumed GBD. The strain energy ΔG_3 is the reduction of the strain energy of GBDs outside the cavity. Full details can be found in [19, 40].

Lim's model has been applied to copper and austenitic stainless. As long as energy is gained when a cavity is formed, that is, ΔG in Eq. (7) is negative, cavitation is possible. From Eq. (7) ΔG is reduced when the applied stress σ_{appl} is raised, that is, cavitation becomes more likely which is natural. On the other hand when σ_{appl} is reduced cavitation is more difficult. There is minimum stress where cavitation is no longer possible because ΔG becomes positive. This minimum stress is plotted as a function of temperature for copper in **Figure 5**.

These minimum stresses are compared with design stresses during creep in copper. It is clear that the stresses required for nucleation are well below the stresses that typically appear in the material. This demonstrates that nucleation based on the substructure is a viable process.

From a technical point of view it is well established that the creep ductility of oxygen free pure copper Cu-OF can be very much lower than for phosphorus alloyed copper Cu-OFP. As a consequence the latter material should be used in creep exposed components [28, 41]. It is evident from **Figure 5** that much lower stresses are needed in Cu-OF than in Cu-OFP, which makes the cavitation in the former material much more abundant. This is believed to be the main reason for the low creep ductility of Cu-OF.

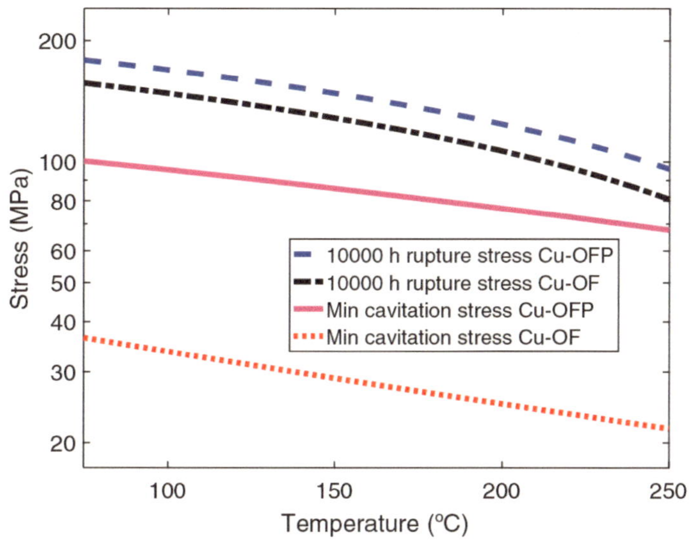

Figure 5. Minimum stress to form cavities at cell boundaries versus temperature for oxygen free pure Cu-OF and phosphorus alloyed copper Cu-OFP. For comparison the stress that gives creep rupture after one year (10,000 h) is included. After Ref. [19].

It has also been verified that the minimum nucleation stresses are below typical design stresses for the common stainless steels 304H (18Cr10Ni), 316 (17Cr12Ni2Mo), 321 (18Cr12NiTi) and 347 (18Cr12NiNb). For example this is illustrated for 347 (18Cr12NiNb) in **Figure 6**. The design stresses are 10,000 h rupture data. The minimum cavitation stress lies in the interval 35–50 MPa in the interval from 500°C to 750°C. The minimum cavitation stresses are again below the design stresses. The temperature dependence of Lim's model is probably not fully correct. In general it is thought that the amount of cavitation will increase with temperature. However the temperature dependence of the minimum cavitation stress is weaker than that of the design stress, which suggests the opposite behaviour.

3.2. Strain dependence

Experimentally it has been found many times that the number of cavities is proportional to the creep strain, cf. Eq. (2). To explain this strain dependence, Sandstrom and Wu introduced the *double ledge model* [43]. They considered a sliding grain boundary with dislocation substructures on both sides of the boundary that moved along with the grains. The substructures consist of subgrains that contain fairly few dislocations in their interior but with well-developed subgrain walls. The positions where the subgrain walls meet at the grain boundary are referred to as subgrain corners. Nucleation was assumed to take place when a subboundary on one side of the boundary hits a subgrain corner on the other side. The nucleation rate can be expressed as

$$\frac{dn}{dt} = \frac{\dot{u}_{\mathrm{GBS}}}{d_{\mathrm{sub}}} \frac{1}{d_{\mathrm{sub}}^2}, \tag{8}$$

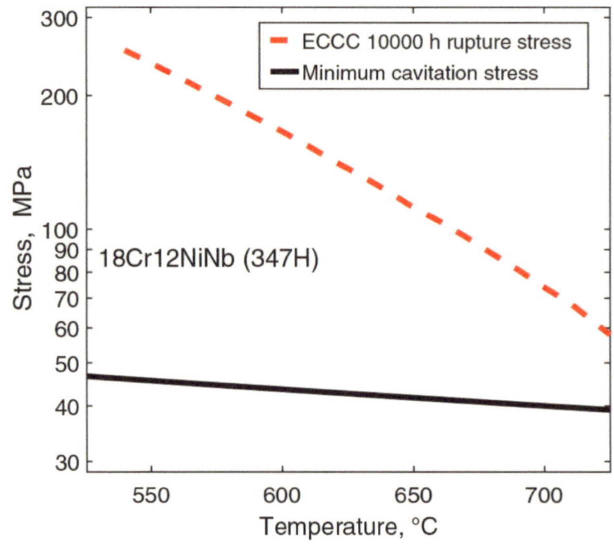

Figure 6. Minimum cavitation stress versus temperature for TP347H austenitic stainless steel. 10,000 h rupture data from ECCC [42] for TP347 are shown for comparison.

where d_{sub} is the subgrain diameter. The last factor takes into account that one nucleus can be formed in each subgrain on the boundary. The subgrain size is directly related to the applied stress [19]

$$d_{sub} = K_{sub} \mathrm{GB}/\sigma_{appl}. \tag{9}$$

The constant K_{sub} is about 20 for austenitic stainless steels and about 11 for copper. The same model can be applied to particles in the grain boundary that are known to contribute to the nucleation. In the model the subgrain corners are replaced by the particles in the grain boundaries with an interparticle distance of λ. Taking both subgrain corners and particles into account, the resulting expression for the nucleation rate is [44]

$$\frac{dn}{dt} = \frac{0.9\, C_s}{d_{sub}}\left(\frac{1}{d_{sub}^2} + \frac{1}{\lambda^2}\right)\dot{\varepsilon} = B\dot{\varepsilon}. \tag{10}$$

In Eq. (10), Eq. (1) has been used. The factor 0.9 in Eq. (10) takes into account the averaging of different orientations [44].

The model in Eq. (10) is compared with experimental data for austenitic stainless steels in **Figure 7**. For three of the experimental data sets TP347 at 550°C and 650°C and TP304 at 727°C, the model gives quite an acceptable description. For TP304XX at 750°C the deviation between model and experiment is larger.

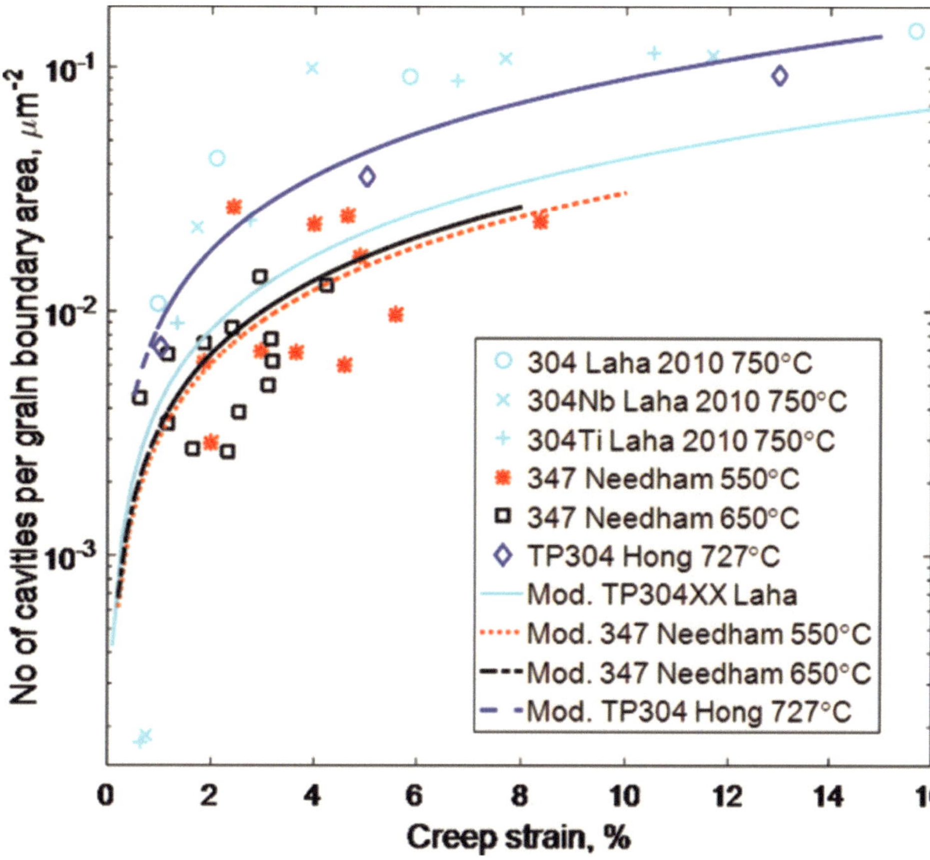

Figure 7. Modelling and experimental number of cavities per unit grain boundary area as a function of creep strain [44]. Experimental data from Hong and Nam [45] for TP304 steel, Laha et al. [46] for three different types of austenitic stainless steels and Needham and Gladman [27] for TP347 steel.

3.3. Particle size

It has been proposed that a critical particle radius exists for nucleation [45, 46]. The radius must exceed a minimum value in order for nucleation to take place. Harris developed a model that related the critical particle size to the GBS velocity [47, 48]. His basic assumption was that particles are not able to stop GBS if the diffusion is fast enough. According to Harris this critical particle radius also represented the minimum radius that could nucleate cavities. Harris gave the following relation between the GBS velocity $\dot{u}_{GBS}$ and the critical particle radius r_c

$$\dot{u}_{GBS} = \frac{\delta D_{GB}}{r_c^2 \ln \frac{\lambda}{2 r_c}} \left(\exp \frac{2 \gamma_s \Omega}{k_B T r_c} - 1 \right), \tag{11}$$

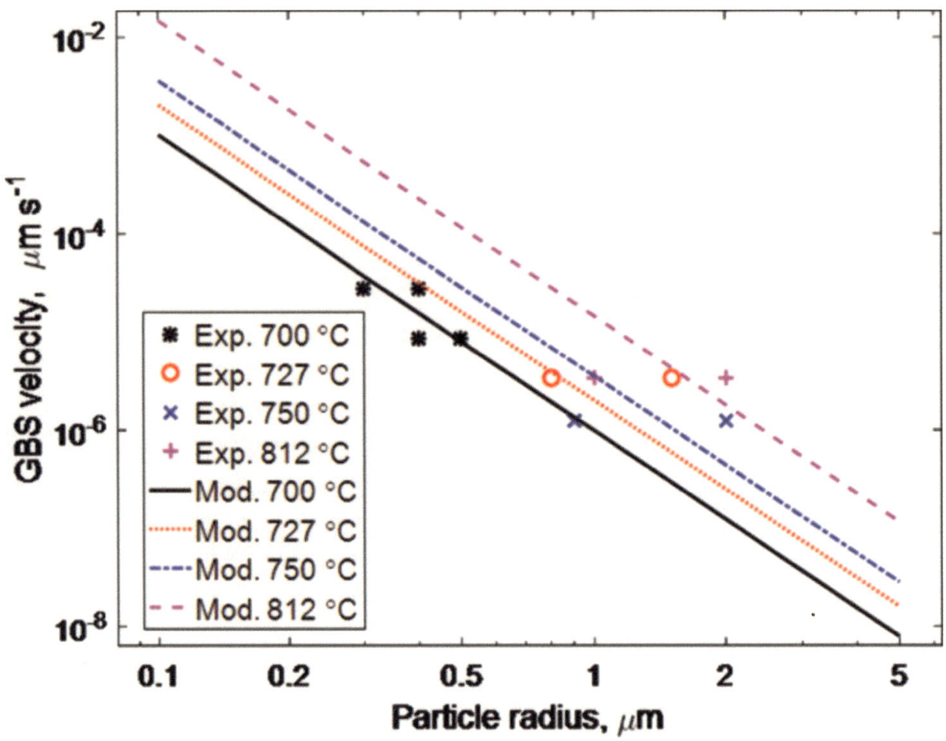

Figure 8. Comparison of experimental and modelling GBS velocity as a function of particle radius [24]. Experimental data for 304, 321 and 347 austenitic stainless steels from Refs. [45, 46, 49, 50]. Modelling results from Eq. (11).

where λ is the interpaticle spacing, γ_s the surface energy, δ the grain boundary width, D_{GB} the grain boundary self-diffusion coefficient, Ω the atomic volume, k_B Boltzmann's constant and T the absolute temperature. The application of Eq. (11) is illustrated in **Figure 8** for different types of austenitic stainless steels [24].

In **Figure 8** the particle parameters are taken from the experimental references. The minimum particles that nucleated cavities in the experiments are chosen for the critical particle radius. The experimental data clearly support Harris' model.

From the particle size distributions [24] the number of nucleated cavities can be estimated if the critical particle size is known. The computed number of nuclei is compared with the observed ones for austenitic stainless steels in **Figure 9.**

4. Cavity growth

4.1. Unconstrained cavity growth model

After the cavities have been nucleated, they start to grow if they exceed a critical size. The main mechanism for the growth is diffusion. Vacancies are transported away from the surfaces of the cavities. The grain boundaries are good sinks for the vacancies. The first model for

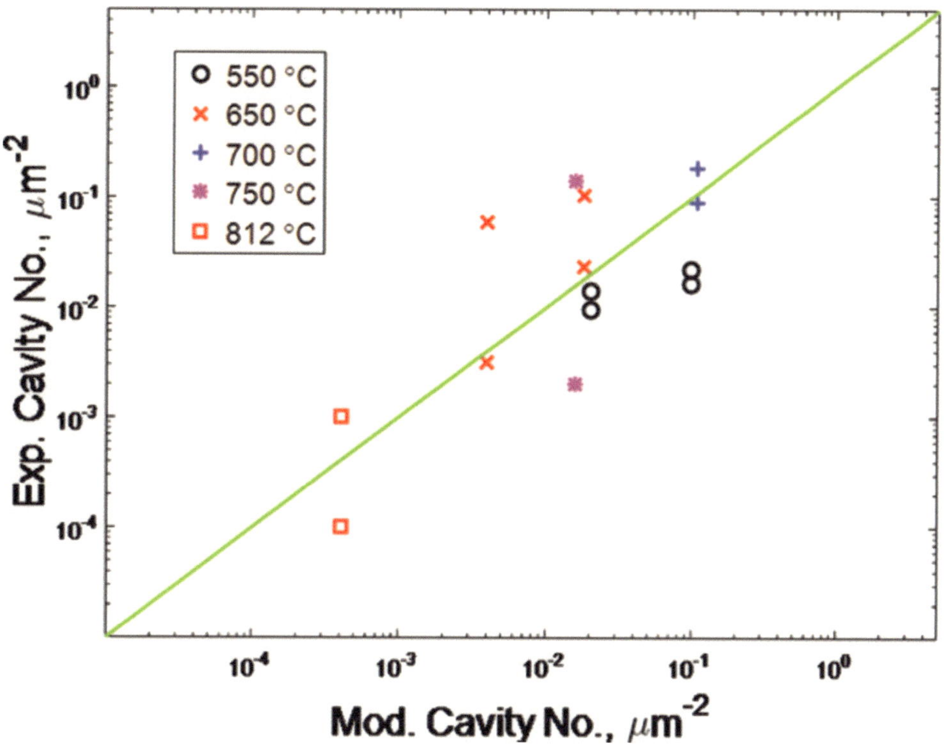

Figure 9. Comparison of experimental number of cavities and modelling number of particles that initiate cavities [24]. Experimental data for austenitic stainless steels from Refs. [27, 45, 46, 49, 50].

diffusion controlled growth was presented by Hull and Rimmer [15]. A much more elegant formulation was later given by Beere and Speight [16] and this is the model that has been used since. Their growth equation can be expressed as

$$\frac{dR}{dt} = 2 D_0 K_f(\sigma_{appl} - \sigma_0) \frac{1}{R^2}, \tag{12}$$

where R the cavity radius in the grain boundary plane, dR/dt its growth rate, σ_0 the sintering stress $2\gamma_s \sin(\alpha)/R$, where γ_s is the surface energy of the cavity per unit area and α the cavity tip angle. D_0 is a grain boundary diffusion parameter, $D_0 = \delta D_{GB}\Omega/k_B T$, where δ is the grain boundary width, D_{GB} the grain boundary self-diffusion coefficient, Ω the atomic volume, k_B Boltzmann's constant and T the absolute temperature. The factor K_f was introduced in [16]. It is a function of the cavitated grain boundary area fraction $f_a = (2R/L)^2$

$$K_f = -1/[2 \log f_a + (1 - f_a)(3 - f_a)]. \tag{13}$$

From the number of cavities per unit grain boundary area n_{cav}, the cavity spacing L can be determined

$$L = 1/\sqrt{n_{cav}}. \tag{14}$$

n_{cav} can be found from the nucleation model, Eq. (10). Plastic deformation can also contribute to the growth rate. Danavan and Solomon have given an expression for that [51]

$$\frac{dR}{dt} = \frac{\sin^2(\alpha)}{\alpha - \sin(\alpha)\cos(\alpha)} \frac{R}{3} \dot{\varepsilon}. \tag{15}$$

4.2. Constrained cavity growth

When diffusion controlled growth models were compared with experimental data, it was evident that the models often strongly exaggerated the growth rate. Dyson found that the predicted growth rate of the cavities many times exceeded the deformation rate of the surrounding material which he considered as unphysical [17]. He suggested that the cavity growth rate should not be larger than the creep rate of the material. This was referred to as constrained growth. Based on this assumption, Rice developed a quantitative model [18]. The result is that in the growth equation, the applied stress is replaced by a reduced stress

$$\frac{dR}{dt} = 2 D_0 K_f(\sigma_{red} - \sigma_0) \frac{1}{R^2}. \tag{16}$$

The reduced stress is given by

$$\sigma_{red} = \sigma_0 + \frac{1}{\frac{1}{\sigma_{appl}} + \frac{32 D_0 K_f}{L^2 d\beta \dot{\varepsilon}(\sigma_{appl})}}, \tag{17}$$

where β is a material constant (β = 1.8 for homogeneous materials) and d the grain diameter. With this approach a growth model that fulfils Dyson's criterion has been achieved.

Rice based his analysis on a linear viscoplastic model of an opening crack. He and Sandstrom reanalyzed the model and avoided the assumption of linearity [52]. A grain structure with a pillar of height h and width corresponding to the grain size d was set up. In this pillar the creep deformation in the axial (z) direction is given by

$$\frac{dz}{dt} = 4\pi D_0 K_f(\sigma_{red} - \sigma_0) n_{cav} + h \dot{\varepsilon}(\sigma_{red}) = h \dot{\varepsilon}(\sigma_{appl}). \tag{18}$$

$\dot{\varepsilon}(\sigma_{red})$ and $\dot{\varepsilon}(\sigma_{appl})$ are the creep rates at the reduced and applied stress, respectively. The first term in the middle part of Eq. (17) is the volume growth rate of a cavity multiplied by the number of cavities per unit grain boundary area. The creep displacement of the pillar at the reduced stress is the second term. The final term on the right hand is the displacement in the surrounding material. A finite element analysis was performed to determine the size of

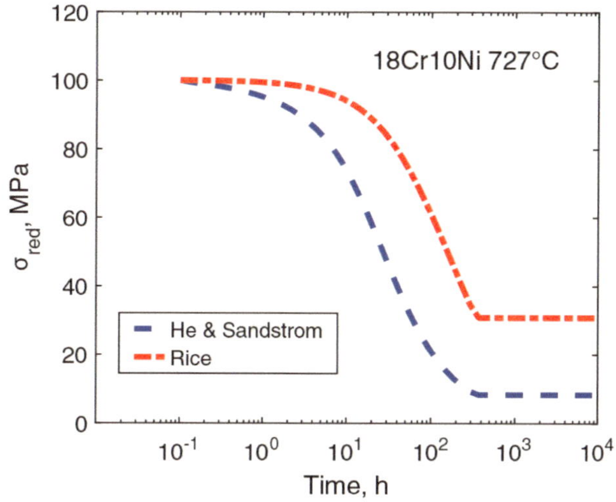

Figure 10. Reduced stress according to Eq. (19) versus time [52]. The result is compared with the model of Rice in Eq. (17) [18]. Cavity growth for 18Cr10Ni at 727°C and 100 MPa [45].

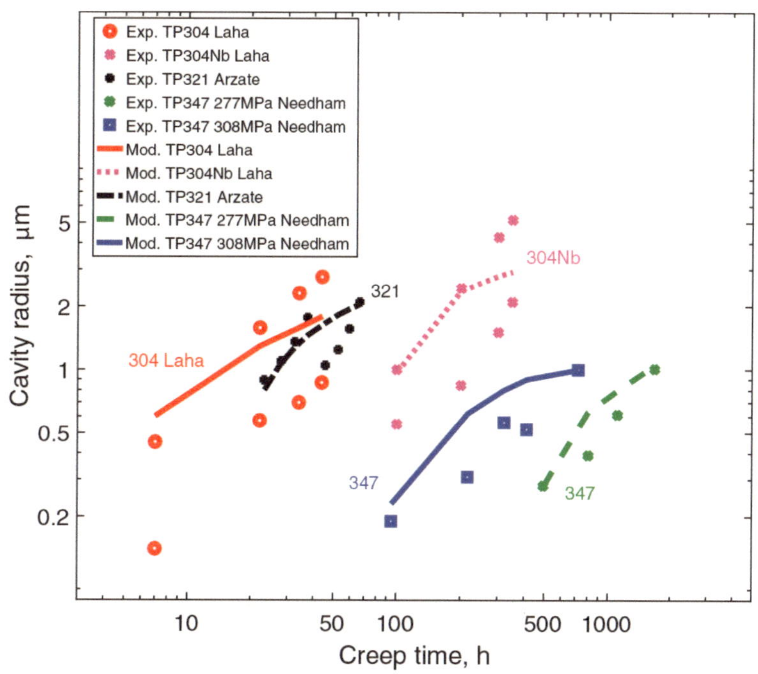

Figure 11. Cavity radius as a function of creep time for 18Cr10Ni without or with Nb (347) or Ti (321) austenitic stainless steels. Model according to Eq. (19) and experimental data from Refs. [27, 46, 49]. The creep tests were performed at temperatures in the interval of 650–812°C.

the height h. It was found that $h \approx 2R$ in the investigated cases [52]. If this value for h is used and n_{cav} is replaced by $1/L^2$ according to Eq. (14), the following equation is obtained

$$2\pi D_0 K_f(\sigma_{red} - \sigma_0) \Big/ L^2 R + \dot{\varepsilon}(\sigma_{red}) = \dot{\varepsilon}(\sigma_{appl}). \tag{19}$$

In general Eq. (19) has to be solved by iteration to find the new value of σ_{red}. This new value for σ_{red} is lower than that given by Eq. (17). This is illustrated in **Figure 10**: Both the absolute and relative difference increase with time.

The new constrained growth model is compared to experimental data for austenitic stainless steels in [52]. Some examples are given here in **Figure 11**. Growth data for 18Cr10Ni steel with and without Nb or Ti are shown. It can be seen that the growth data can be described with fair accuracy. The lower growth rate according Eq. (19) is important in this respect.

5. Brittle creep rupture

Creep rupture is technically very important, because it determines the life of many plants operating at high temperatures. Two main mechanisms are distinguished: ductile rupture and brittle rupture. Ductile rupture is controlled by the exhaustion of the deformation capacity of the material. In this case the usual rupture criterion is that the creep strain reaches a critical value. The deformation takes place by dislocation mechanisms. The faster the dislocations move, the faster rupture occurs. Since ductile rupture does not involve cavitation, it is not reviewed here, but full details can be found elsewhere [53].

Rupture curves for dislocation creep are illustrated in **Figure 12** for the austenitic stainless steel 18Cr12NiTi (321H) at temperatures between 600°C and 775°C. The experimental creep rupture data cover times up to 100,000 h (11 years). The general overall behaviour is well described by the model predictions.

The second process brittle rupture is due to grain boundary decohesion. By far the most important mechanism in this respect is the formation and growth of cavities. It is well established that when the cavitated grain boundary area reaches a certain fraction of about 0.25, brittle rupture takes place [54]. The cavitated area fraction A_{cav} can be computed from Ref. [43]

$$A_{cav} = \int_{t_i}^{t} \frac{dn}{dt'}(t')\pi R^2(t, t')dt'. \tag{20}$$

A continuous nucleation of cavities takes place. The number of cavities is directly proportional to the creep strain, Eq. (10). Once a cavity has nucleated it starts to grow after an incubation time t_i that is a small fraction of the rupture time [52]. The growth is described with Eq. (16) with the reduced stress given by Eq. (19). When A_{cav} has reached 0.25, rupture is assumed to take place.

The model predictions for brittle rupture for 18Cr12NiTi (321H) are shown in **Figure 13**. The predictions are compared to the same experimental data as in **Figure 12**. Again the overall

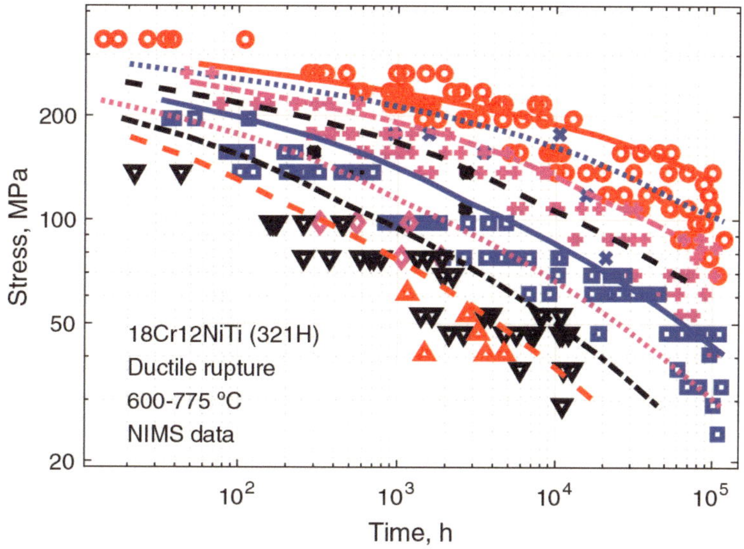

Figure 12. Comparison of dislocation creep model rupture curves (ductile rupture) Refs. [53, 54] with experiments [55] for 18Cr12NiTi (321H). Model prediction and experiments at temperatures between 600°C and 775°C with 25°C interval.

time dependence of the rupture strength at different temperatures is well represented. In fact the differences between the model predictions for ductile rupture in **Figure 12** and brittle rupture **Figure 13** are not very large.

Ductile rupture is assumed to be controlling if the strain exhaustion occurs before $A_{cav} = 0.25$ has been reached. On the other hand if the cavitation criterion is reached first, brittle rupture takes place. The results for ductile and brittle rupture are combined in **Figure 14**. For a given

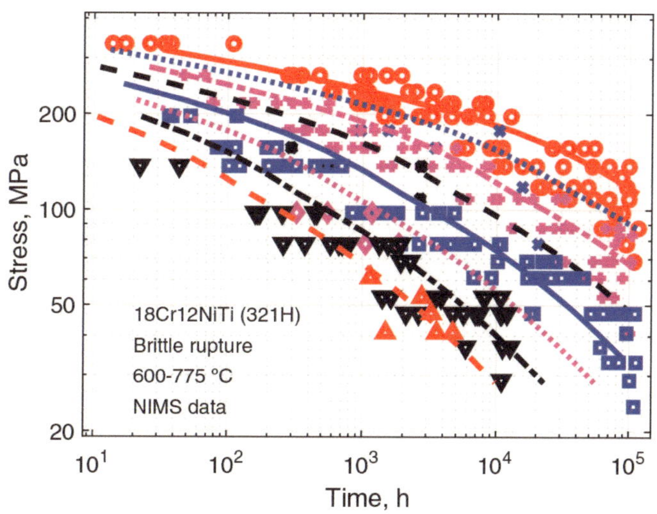

Figure 13. Comparison of model rupture curves based on cavitation (Eq. (20), brittle rupture) [54] with experiments [55] for 18Cr12NiTi (321H). Model prediction and experiments at temperatures between 600°C and 775°C with 25°C interval.

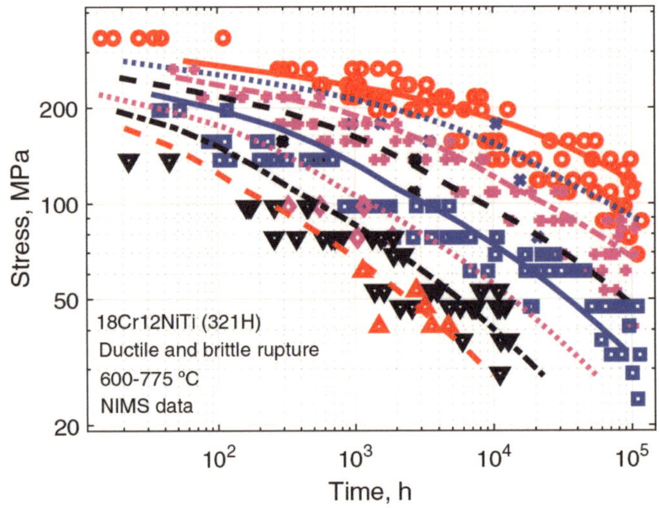

Figure 14. Comparison of model rupture curves based on both dislocation creep (ductile rupture) and cavitation (brittle rupture) with experiments [55] for 18Cr12NiTi (321H). Model predictions and experiments at temperatures between 600°C and 775°C with 25°C interval.

temperature and stress the value from **Figure 12** is chosen if the (ductile) rupture time is shorter than the (brittle) rupture time in **Figure 13** and vice versa.

When brittle rupture is taken into account when modelling the creep rupture curves, there is an improvement in particular at high temperatures and low stresses.

6. Discussion

With the development of the *shear sliding* and *shear crack models* for grain boundary sliding (GBS), it is now possible to compute the displacements across grain boundaries in a quantitative way and the results are in acceptable agreement with experiments. This has given a dramatic improvement in the understanding of GBS. The previous observations that the amount of grain boundary sliding is proportional to the creep strain are reproduced by the model. According to the shear sliding model the displacement is proportional to the grain size: This has been verified experimentally for small grain sizes, but the grain size dependence is probably exaggerated for larger grain sizes. A detailed comparison between the shear crack model and experiments is more difficult, since in none of the published results on GBS, full details of the particle structure are presented.

It is assumed in general that nucleation of creep cavities is based on GBS. The new models for GBS have made it possible to set up realistic models for nucleation. Many models for nucleation developed in the past are strongly stress dependent and suggest that the amount of cavitation would increase with stress, which is clearly at variance with most observations on creep. This applies for example to models based on classical nucleation theory. Using Lim's model for substructure induced cavitation [40], it has now been demonstrated both for commercial copper and austenitic stainless steel that the creep stresses in the substructure are sufficiently high to nucleate cavities. Since the stresses are stationary, the problem with rapid relaxation in particular at high stresses is avoided. If particles instead of subgrain corners constrain the grain boundary dislocations, the outcome of the model is essentially the same. The model is consequently applicable to cavity nucleation at particles as well.

With the help of the *double ledge model* it can be explained why the nucleation rate is proportional to the creep strain rate. According to the model nucleation takes place when subgrain boundaries meet subgrain corners on the other side of a sliding grain boundary. The displacement rate is proportional to creep strain rate according to the GBS models. As a consequence the suboundaries will meet the subgrain corners at the same rate and this explains why the nucleation rate is proportional to the strain rate.

Harris developed a model for nucleation around particles many years ago [47, 48]. He considered the relation between particle sizes and the GBS displacement. If the particles are sufficiently large it is assumed that they will prevent GBS. He formulated a criterion for the critical particle size. If GBS is prevented significant stresses are formed at the particles. If the critical particle size is exceeded, Harris proposed that cavity nucleation can take place. With the event of the new models for GBS it has now for the first been possible to test Harris' ideas. It turns out that both the critical particle radius and the number of cavities that can be estimated from the particle distributions are in agreement with observations for austenitic stainless steels. In this way a method for estimating the critical particle size for nucleation has been established.

GBS does not occur on every grain boundary. In fact only on a limited number of GBS events is observed in a material that has been creep exposed [19]. As a consequence cavity nucleation only occurs on some grain boundaries. The presence of cavities show a large statistical variation, see for example [39]. The models in the present paper represent grain boundaries where cavities will be formed. These are also the grain boundaries that control the rupture of the material. The models predict a fairly high nucleation rate that represents the most active grain boundaries and not an average over all grain boundaries.

Models for cavity growth have been available for a long time. Unfortunately these models in general generate growth rates that are much higher than the observed ones. This might be the reason why very few quantitative comparisons have been made between experiments and models in the literature. A major step forward was the introduction of constrained growth [17]. Then the cavities were not allowed to grow faster than the surrounding creeping material. A quantitative model for constrained growth was given by Rice [18]. Still the growth rates tended to be higher than the observed ones. Only recently it has been recognised that some minor approximations in the work of Rice were essential to correct [52]. With these new corrections satisfactory predictions for cavity growths in austenitic stainless have been achieved.

The recent development has implied that quantitative models are now available for grain boundary sliding, for the thermodynamic feasibility of cavity nucleation controlled by subboundaries, the nucleation rate and constrained cavity growth, that is, for all of the involved main processes. Each model is of great significance in itself, but by using combinations of them, the development of cavities can be fully understood.

7. Conclusions

1. Recent development of processes concerning the development of creep cavities has been reviewed. Models have been covered for grain boundary sliding, cavity nucleation and cavity growth.

2. Based on analysis of previously presented FEM models, a model for the grain boundary displacement during sliding called the shear sliding model has been presented. The model gives a displacement that is proportional to the creep strain. The model can quantitatively reproduce the observed displacements for fcc alloys.

3. Using Lim's model for subboundary assisted cavity nucleation, it has been demonstrated that this process is thermodynamically feasible for copper and for austenitic stainless steels. It gives a minimum cavitation stress that is well below stresses in creep exposed components.

4. According to the double ledge model, nucleation is assumed to take place when subboundaries on one side of a sliding grain boundary meet subgrain corners or particles on the other side. The model gives a nucleation rate that is proportional to the creep rate in good accordance with observations. The model can predict the measured nucleation rates in austenitic stainless steels.

5. Due to the development of the new models for GBS, it has for the first time been possible to test Harris' model for cavity nucleation around particles. The model gives a relation between the critical particle radius and the GBS velocity. Data for austenitic stainless steels confirm the validity of the model.

6. A modified constrained growth model has been presented. It gives lower growth rates than previous models. These lower growth rates have implied that observed growth rates for austenitic stainless steels can now be reproduced.

7. Alloys can fail by ductile or brittle creep rupture. It is demonstrated that by considering both ductile and brittle rupture, the prediction of creep rupture curves can be improved in particular at high temperatures and low stresses.

Author details

Rolf Sandström* and Junjing He

*Address all correspondence to: rsand@kth.se

Materials Science and Engineering, KTH Royal Institute of Technology, Stockholm, Sweden

References

[1] B. Neubauer, Recent Advances in Creep and Fracture of Engineering Materials and Structures, B. Wilshire, D.R.J. Owen (eds.). Pineridge Press, Swansea, 1981, 617–619.

[2] B. Neubauer, F. Arens-Fischer, Determination of residual life of power station components subject to creep stress, VGB Kraftwerkstechnik, 63 (1983) 637–645.

[3] L.M. Kachanow, Izvestiya Akademii Nauk SSSR, Otdelenie Tekhnicheskikh Nauk, Mekhanika i Mashinostroenie, 8 (1958), 26–31.

[4] Kachanow, L.M., Foundations of Fracture Mechanics. Moscow, 1974.

[5] Rabotnov, Y.N., Creep Problems in Structural Members. NorthHolland, Amsterdam, 1969.

[6] R. Sandstrom, A. Kondyr, Model for Tertiary-Creep in Mo and CrMo-Steels, in: Proceedings—Computer Networking Symposium, 1980, 275–284.

[7] R. Sandstrom, A. Kondyr, Creep Deformation, Accumulation of Creep Rupture Damage and Forecasting of Residual Life for Three Mo- and CrMo-Steels, VGB Kraftwerkstechnik, 62 (1982), 802–813.

[8] R. Sandstrom, Basic Model for Primary and Secondary Creep in Copper, Acta Materialia, 60 (2012) 314–322.

[9] M.H. Yoo, H. Trinkaus, Crack and Cavity Nucleation at Interfaces During Creep, Metallurgical Transactions A, 14 (1983) 547–561, doi:10.1007/BF02643772

[10] R. Raj, M.F. Ashby, Intergranular Fracture at Elevated Temperature, Acta Metallurgica, 23 (1975) 653–666.

[11] E. Smith, J.T. Barnby, Crack Nucleation in Crystalline Solids, Metal Science, 1 (1967) 56–64.

[12] R. Raj, Nucleation of Cavities at Second Phase Particles in Grain Boundaries, Acta Metallurgica, 26 (1978) 995–1006.

[13] H. Riedel, Life Prediction Methods for Constrained Grain Boundary Cavitation, International Journal of Pressure Vessels and Piping, 39 (1989) 119–134.

[14] X.G. Jiang, J.C. Earthman, F.A. Mohamed, Cavitation and Cavity-Induced Fracture During Superplastic Deformation, Journal of Materials Science, 29 (1994) 5499–5514.

[15] D. Hull, D.E. Rimmer, The Growth of Grain Boundary Voids Under Stress, Philosophical Magazine, 4 (1959) 673–687.

[16] W. Beere, M.V. Speight, Creep Cavitation by Vacancy Diffusion in Plastically Deforming Solid, Metal Science, 21 (1978) 172–176.

[17] B.F. Dyson, Constraints on Diffusional Cavity Growth Rates, Metal Science, 17 (1976) 349–353.

[18] J.R. Rice, Constraints on the Diffusive Cavitation of Isolated Grain Boundary Facets in Creeping Polycrystals, Acta Metallurgica, 29 (1981) 675–681.

[19] R. Sandström, R. Wu, J. Hagström, Grain Boundary Sliding in Copper and its Relation to Cavity Formation During Creep, Materials Science and Engineering A, 651 (2016) 259–268.

[20] C.W. Chen, E.S. Machlin, Trans. AIME J. Metals, 209 (1957) 829–835.

[21] J. Intrater, E.S. Machlin, J. Inst. Metals, 88 (1959/60) 305–310.

[22] B.J. Cane, Mechanistic Control Regimes for Intergranular Cavity Growth in 2. 25Cr-1Mo Steel Under Various Stresses and Stress States, Metal Science, 15 (1981) 302–310.

[23] D.J. Gooch, Creep Fracture of 12Cr-Mo-V Steel, Metal Science, 16 (1982) 79–89.

[24] J. He, R. Sandström, Modelling Grain Boundary Sliding During Creep of Austenitic Stainless Steels, Journal of Materials Science, 51 (2016) 2926–2934.

[25] D. McLean, M.H. Farmer, The Relation During Creep Between Grain-Boundary Sliding, Sub-Crystal Size and Extension, Journal of the Institute of Metals, 85 (1957) 41–50.

[26] N.G. Needham, J.E. Wheatley, G.W. Greenwood, The Creep Fracture of Copper and Magnesium, Acta Metallurgica, 23 (1975) 23–27.

[27] N.G. Needham, T. Gladman, Nucleation and Growth of Creep Cavities in a Type 347 Steel, Metal Science, 14 (1980) 64–72.

[28] R. Wu, R. Sandstrom, Strain Dependence of Creep Cavity Nucleation in Low Alloy and 12%Cr Steels, Journal of Materials Science & Technology, 12 (1996) 405–415.

[29] K. Pettersson, A Study of Grain Boundary Sliding in Copper With and Without an Addition of Phosphorus, Journal of Nuclear Materials, 405 (2010) 131–137.

[30] A. Ayensu, T.G. Langdon, The Inter-Relationship Between Grain Boundary Sliding and Cavitation During Creep of Polycrystalline Copper, Metallurgical and Materials Transactions A: Physical Metallurgy and Materials Science, 27 (1996) 901–907.

[31] F.W. Crossman, M.F. Ashby, The Non-Uniform Flow of Polycrystals by Grain-Boundary Sliding Accommodated by Power-Law Creep, Acta Metallurgica, 23 (1975) 425–440.

[32] F. Ghahremani, Effect of Grain Boundary Sliding on Steady Creep of Polycrystals, International Journal of Solids and Structures, 16 (1980) 847–862.

[33] H. Riedel, Cavity Nucleation at Particles on Sliding Grain Boundaries. A Shear Crack Model for Grain Boundary Sliding in Creeping Polycrystals, Acta Metallurgica, 32 (1984) 313–321.

[34] Laha K, Kyono J, Sasaki T, Kishimoto S, Shinya N Improved Creep Strength and Creep Ductility of Type 347 Austenitic Stainless Steel Through the Self-Healing Effect of Boron for Creep Cavitation, Metallurgical and Materials Transactions A, 36A (2005) 399–409.

[35] Kishimoto S, Shinya N, Tanaka H, Grain Boundary Sliding and Surface Cracking during Creep of 321 Stainless Steel, Materials, 37 (414) (1987) 289–294.

[36] Gates RS, Stevens RN, The Measurement of Grain Boundary Sliding in Polycrystals, Metallurgical Transactions, 5 (1974) 505–510.

[37] Gittins A, The Kinetics of Cavity Growth in 20 Cr25 Ni Stainless Steel, Journal of Materials Science, 5 (1970) 223–232.

[38] Morris DG, Harries DR, Wedge Crack Nucleation in Type 316 Stainless Steel, Journal of Materials Science, 12 (1977) 1587–1597.

[39] T.G. Langdon, The Role of Grain Boundaries in High Temperature Deformation, Materials Science and Engineering A, 166 (1993) 67–79.

[40] L.C. Lim, Cavity Nucleation at High Temperatures Involving Pile-ups of Grain Boundary Dislocations, Acta Metallurgica, 35 (1987) 1663–1673.

[41] P.J. Henderson, R. Sandstrom, Low Temperature Creep Ductility of OFHC Copper, Materials Science and Engineering A, 246 (1998) 143–150.

[42] ECCC European Creep Collaborative Committee Data Sheets 2005—HR3C.120, 2005.

[43] R. Sandström, R. Wu, Influence of Phosphorus on the Creep Ductility of Copper, Journal of Nuclear Materials, 441 (2013) 364–371.

[44] J. He, R. Sandström, Formation of Creep Cavities in Austenitic Stainless Steels, Journal of Materials Science, 51 (2016) 6674–6685.

[45] Hong JH, Nam SW, Choi SP, The Influences of Sulphur and Phosphorus Additions on the Creep Cavitation Characteristics in Type 304 Stainless Steels, Journal of Materials Science 21 (11) 3966–3976.

[46] Laha K, Kyono J, Shinya N, Suppression of Creep Cavitation in Precipitation-Hardened Austenitic Stainless Steel to Enhance Creep Rupture Strength. Transactions of the Indian Institute of Metals, 63 (2–3) (2010) 437–441.

[47] J.E. Harris, An Analysis of Creep Ductility of Magnox Al80 and its Implications, Journal of Nuclear Materials, 15 (1965) 201–207.

[48] J.E. Harris, Nucleation of Creep Cavities in Magnesium, Transactions of the Metallurgical Society of AIME, 233 (1965) 1509.

[49] O.R. Arzate, L. Martinez, Creep Cavitation in Type 321 Stainless Steel, Material Science and Engineering A, 101 (1988) 1–6.

[50] K. Laha, J. Kyono, T. Sasaki, S. Kishimoto, N. Shinya, Improved Creep Strength and Creep Ductility of Type 347 Austenitic Stainless Steel through the Self-Healing Effect of Boron for Creep Cavitation, Metallurgical and Materials Transactions A, 36A (2005) 399-409.

[51] K. Davanas, A.A. Solomon, Theory of Intergranular Creep Cavity Nucleation, Growth and Interaction, Acta Metallurgica et Materialia, 38 (1990) 1905–1916.

[52] J. He, R. Sandström, Creep Cavity Growth Models for Austenitic Stainless Steels, Materials Science and Engineering A, 674 (2016) 328–334.

[53] S. Vujic, R. Sandstrom, C. Sommitsch, Precipitation Evolution and Creep Strength Modelling of 25Cr20NiNbN Austenitic Steel, Materials at High Temperatures, 32 (2015) 607–618.

[54] J. He, R. Sandström, Basic Modelling of Creep Rupture in Austenitic Stainless Steels, to be publ. (2017).

[55] NRIM, National Research Institute for Metals, Tokyo, Japan, 1987.

Chapter 3

Formation Mechanism and Elimination of Small-Angle Grain Boundary in AlN Grown on (0001) Sapphire Substrate

Ryan G. Banal, Masataka Imura and Yasuo Koide

Additional information is available at the end of the chapter

http://dx.doi.org/10.5772/66177

Abstract

AlN epilayers were grown on (0001) sapphire substrates by metal-organic vapor phase epitaxy (MOVPE), and the influence of substrate's surface structure on the formation of in-plane rotation domain is studied. The surface structure of sapphire substrate is found to change during thermal cleaning and temperature ramp-up. The oxygen-terminated surface with monolayer (ML) steps having *ABAB* stacking is attributed to cause the formation of a small-angle grain boundary (SAGB). To suppress the formation of such a special grain boundary, the two-step temperature growth technique was employed. It was found that a proper timing of the low-temperature buffer layer (LT BL) growth is necessary to eliminate the SAGB. Moreover, the BL growth temperature (T_g) is also found to affect the surface morphology and structural quality of AlN epilayer. The optimized LT BL T_g is found to be 1050°C. The optical emission property by cathodo-luminescence (CL) measurements showed higher emission intensity from AlN epilayer without SAGB.

Keywords: small-angle grain boundary, SAGB, AlN, sapphire substrate, MOVPE, cathodoluminescence, deep-UV, two-step temperature growth

1. Introduction

Single crystal sapphire with corundum structure is a widely used substrate for film deposition and epitaxial growth in many technological fields, such as in optoelectronics for the growth of AlN, GaN, and InN nitride materials. This material exhibits high melting point (2050°C), extremely high chemical stability even at high temperatures, and transparency in the ultraviolet (UV) region, making it a suitable substrate especially for the growth of AlN, which requires high temperature above 1200°C due to the high viscosity of Al atoms. Sapphire also exhibits a hardness of 9 in the Mohs scale, compared to 10 for diamond. On the other hand, AlN is a promising material for UV and deep-UV light emitters and power electronic devices because of its wide bandgap energy (6.05 eV), good stability at elevated temperature, high thermal conductivity (3.4 $W \cdot cm^{-1} \cdot K^{-1}$) and high electric breakdown field (11.7×10^6 $V \cdot cm^{-1}$). Although the native bulk AlN or GaN substrates are already available for homoepitaxial growths, the utilization of sapphire as the substrate material for heteroepitaxial growth of AlN, GaN, InN, and other emerging materials is expected for the years to come, owing to its mature growth technology, availability of large size wafer, and cost advantage [1–2]. In fact, the advances in heteroepitaxial growths have already successfully demonstrated deep-UV light-emitting diodes (LEDs) and photo-pumped AlGaN multi-quantum well lasers [3–7]. However, the radiative emission efficiencies of deep-UV light emitters are still low, prompting for further reduction of dislocations that act as nonradiative recombination centers [3–6].

The heteroepitaxial growth of AlN on sapphire substrate induces several types of dislocations that are driven by their lattice mismatch and difference in crystal structure. With lattice mismatch, a pseudomorphic growth initially occurs, followed by misfit dislocations after exceeding the critical thickness for plastic relaxation. A 30° rotation of AlN epilayer with respect to sapphire substrate in the basal (0001) plane occurs [8]. However, the development of various growth methods has improved the epitaxial quality of AlN in recent years. These growth methods include alternating supply of source precursors (e.g., modified migration-enhanced epitaxy (MEE)) [9–15], direct and high-temperature growth [16–18], substrate pretreatment (e.g., nitridation) [19, 20], two-step low-temperature (LT) AlN buffer layer (BL) and high-temperature (HT) growth [15, 21–23], multiple-step V/III growth [18, 24], precursor preflow [25], and so on. However, despite the improvement in the surface morphology and structural quality of AlN epilayer, the existence of in-plane rotation domain as exhibited by small-angle grain boundary (SAGB) is still observed, regardless of growth method employed [16, 19–21, 25, 26]. This kind of defect must be eliminated as it can have a negative impact in the optical as well as electrical properties of the devices by acting as barriers for transport or carrier sinks. Small-angle grain boundary is one type of special grain boundary which results when the two crystals have only a slight misorientation relative to each another. Moreover, this kind of special grain can be characterized as pure low-angle tilt boundary or pure low-angle twist boundary, where the former is composed of an array of parallel edge dislocations, while the latter is characterized as the slight rotation of crystals about a common axis which is normal to the plane of the boundary.

2. Origin of small-angle grain boundary (SAGB)

The SAGB in AlN grown on (0001) sapphire substrate is considered to originate from the substrate's surface structure. As in any heteroepitaxial growth, the surface structure influences the growth mode. For example, the appearance of a defect structure on the substrate surface (e.g., protrusion) could possibly lead to spiral growth. It is therefore necessary to keep the surface free from any defects as possible. However, as-received sapphire substrates are not free from any surface defects even after undergoing polishing treatment. This includes scratches on the surface, as shown in **Figure 1(a)**. Hence, thermal cleaning under H_2 ambient is performed prior to AlN growth either at the same AlN growth temperature or slightly above it. After thermal cleaning, the substrate's surface transformed into parallel step-and-terrace structure as shown in **Figure 1(b)**. The estimated step height from the line scan is about 0.21 nm.

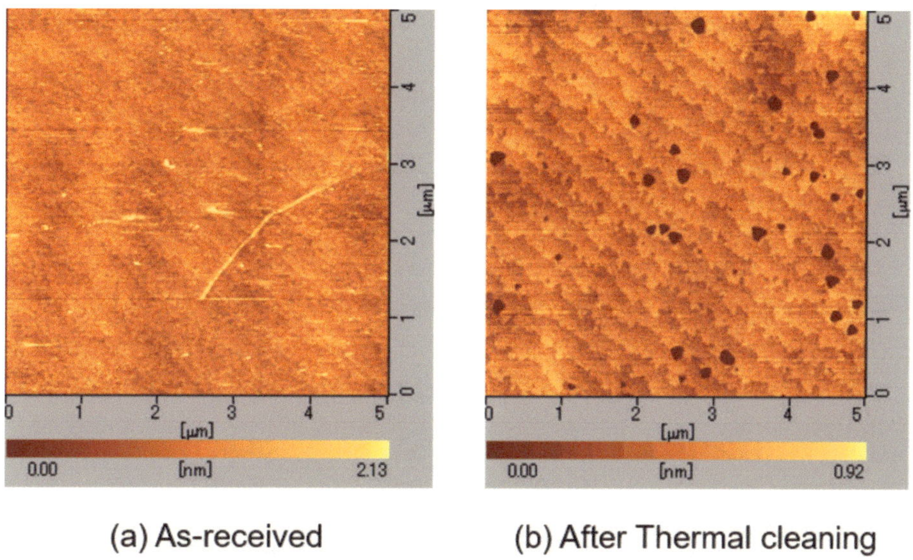

Figure 1. AFM surface morphology of (a) as-received and (b) after thermally cleaned sapphire substrate under H_2 ambient (Ph.D. Thesis, R.G. Banal, Kyoto University).

The crystal lattice of sapphire (α-Al_2O_3) is formed by Al^{3+} and O^{2-} ions. In Al_2O_3 corundum structure, O^{2-} ions are shifted slightly from the idealized hexagonal close-packed positions within the (0001) basal plane due to the empty octahedral sites (note that only two out of every three octahedral sites are occupied by Al^{3+} cations) as shown in **Figure 2(a)** [27, 28]. This results in the formation of two distorted oxygen hexagonal layers as also indicated in **Figure 2(a)** appearing alternately along the [0001] direction with monolayer (ML) periodicity (**Figure 2(b)**). The two distorted oxygen hexagonal layers are labeled as *A* and *B* stacking. By adapting the Thompson's notation [29], the sense of rotation of the distorted hexagon for each oxygen layer can be determined (**Figure 2(c)**). Hence, the successive oxygen layers with *AB* stacking (one ML step) create an opposing rotation, either inwardly or outwardly (**Figure 2(d)**), while the *AA*(*BB*) oxygen stacking (two ML step) would have the same rotation direction either clockwise or counter-clockwise [26, 29]. On the other hand, the coulombic repulsion between Al^{3+} cation causes each to move slightly toward the adjacent unoccupied octahedral site along the [0001] direction (perpendicular to the (0001) basal plane). This results in the formation of slightly puckered layer of Al in the basal plane, where it follows a face-centered cubic-type *abc* stacking. Taking into account the periodic spacing of both the cation and anion layers, the structure repeats itself after six oxygen layers and six double layers of Al^{3+} cation (= 0.1299 nm) [28]. Therefore, the step height between the *A* and *B* oxygen stacking is equal to 0.217 nm as the monolayer step (**Figure 2(b)**). Moreover, no such opposing in-plane rotational geometry can be deduced from the successive Al hexagon layers in contrast to that of the distorted oxygen hexagons, suggesting that the origin of SAGB comes from the oxygen-terminated surface of sapphire substrate.

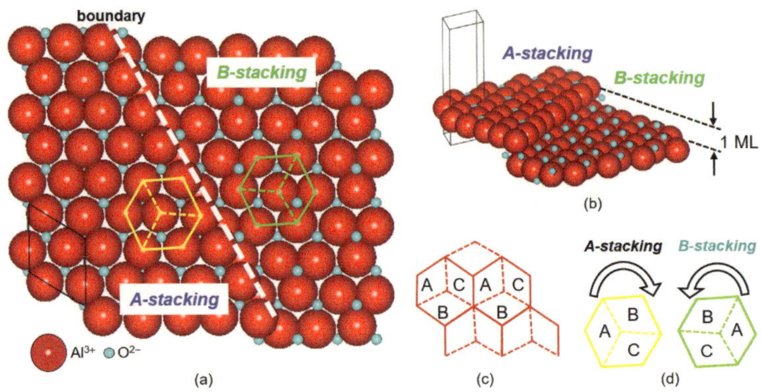

Figure 2. (a) Schematic of distorted oxygen hexagon layers in *A* and *B* stacking. (b) Side view of the *A* and *B* oxygen stacking showing the one monolayer step height. (c) Adaptation of the Thompson's notation to identify the rotation of the distorted oxygen hexagons. (d) Identification of rotation of distorted hexagons from *A* and *B* oxygen stacking layer.

To confirm this hypothesis, the AlN epilayer was grown directly on thermally cleaned sapphire substrate, which is having a monolayer step-and-terrace structure. The AlN growth temperature was 1285°C and the AlN thickness was about 0.9 μm. The atomic force microscopy (AFM) surface morphology of AlN (**Figure 3(a)**) indicates step-and-terrace structure which replicates the surface of thermally cleaned sapphire substrate. The x-ray diffraction

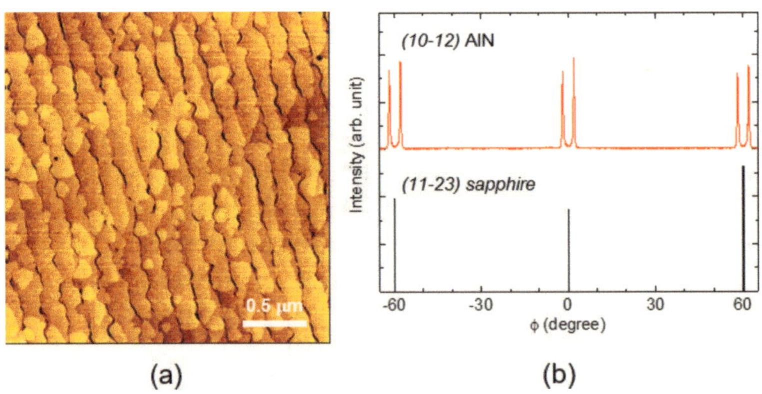

Figure 3. (a) AFM surface morphology of AlN grown directly on thermally annealed sapphire substrate (T_g = 1285°C). (b) (10-12) XRD ϕ-scan of AlN showing twin peaks which correspond to two AlN grains rotated in the in-plane. The (11-23) phi-scan of the sapphire substrate is also shown.

(XRD) ϕ-scan measurement of the AlN (10-12) asymmetric plane observed two peaks, which is attributed to the two periodic grains in AlN, as shown in **Figure 3(b)**. Moreover, because the XRD ϕ-scan did not show any double domain structure from the sapphire substrate, the periodic domain is only observed in the AlN epilayer. In the previous study using similar growth method, cross-sectional transmission electron microscope (TEM) analysis confirmed the existence of two grains as indicated by their periodic bright and dark contrast [26]. The plan-view high-resolution TEM observation further shows an array of edge dislocations along the grain boundary between the two AlN grains [26]. Hence, the monolayer step *ABAB* oxygen stacking is likely the origin of the small-angle grain boundary. Therefore, the surface structure of the sapphire substrate must be prevented from having such structure or modified in order to effectively eliminate the SAGB.

Several techniques have been introduced to eliminate the rotation domain. These include pre-nitrogen radical treatment of the nitrided sapphire substrate [19] and post-annealing after AlN growth [20]. However, these techniques not only entail an additional process but also obtain unsatisfactory results. Another technique is by thermal annealing in the air of sapphire substrate [26]. With the proper annealing temperature and off-cut angle, a substrate surface with *AA*(*BB*) stacking and two monolayer step height can be achieved through step bunching. On the other hand, while the TMA preflow [25] and the LT-AlN BL with pre-nitridation [21] seem promising in situ methods to eliminate the small-angle grain boundary, the influence of substrate's surface structure on its formation/elimination is not yet investigated in detail. Hence, the fragmentary understanding of the influence of surface structure is also evident after the substrate is being subjected to thermal cleaning prior to AlN growth [11, 14, 16, 18, 21, 24] or the lack of it [10, 13, 15, 22, 25]. Therefore, in this chapter, we study the influence of surface structure of sapphire substrate on the formation/elimination of SAGB to improve the quality of AlN epilayer. Then, we introduce the low-temperature (LT) AlN buffer layer technique, with emphasis on its proper timing, to eliminate the SAGB.

3. Experimental methodology

The AlN epilayers were grown on (0001) sapphire substrate by metal-organic vapor phase epitaxy. Trimethylaluminum (TMA) and NH_3 were used as source precursors for Al and N, respectively, while H_2 was used as the carrier gas. The total reactor pressure was kept at ~12 Torr. During the LT-AlN BL growth, the NH_3 and TMA flowrates were set to 1000 and 55 sccm, respectively; while, during HT-AlN growth, the NH_3 and TMA flowrates were set to 130 and 45 sccm, respectively. To study the influence of substrate's surface on the structural as well as optical quality of AlN, the temperature profiles depicted in **Figure 4** (*Profile a* and *Profile b*) were employed. For AlN growth under *Profile a*, the substrate thermal cleaning was introduced for 10 min under H_2 ambient at the same optimized growth temperature for HT-AlN (T_g =1285°C). Then the temperature was lowered to 1100°C for the growth of ~15-nm-thick LT-AlN BL. The temperature was then increased to 1285°C for the growth of ~1-μm-thick HT-AlN. For growth under *Profile b*, thermal cleaning was not introduced. Rather, the temperature was immediately brought to the desired BL T_g (800-1100°C) for the growth of

~15-nm-thick LT-AlN BL. Then the temperature was increased to T_g = 1285°C for HT-AlN growth. In the experiment, all temperature readings are from those indicated by the thermocouple placed near the substrate. Note that although both profiles incorporate LT-AlN BL, the timing, hence, the substrate's surface structure at which the LT-AlN BL is introduced is quite different, which is crucial for the formation/elimination of small-angle grain boundary. For analyses, atomic force microscopy (AFM) measurements were conducted to study the surface morphologies both of the substrate's surface and AlN epilayer, while XRD and transmission electron microscope (TEM) measurements were conducted to study the structural qualities and to assess the SAGB. CL measurements were conducted to study the optical properties of AlN.

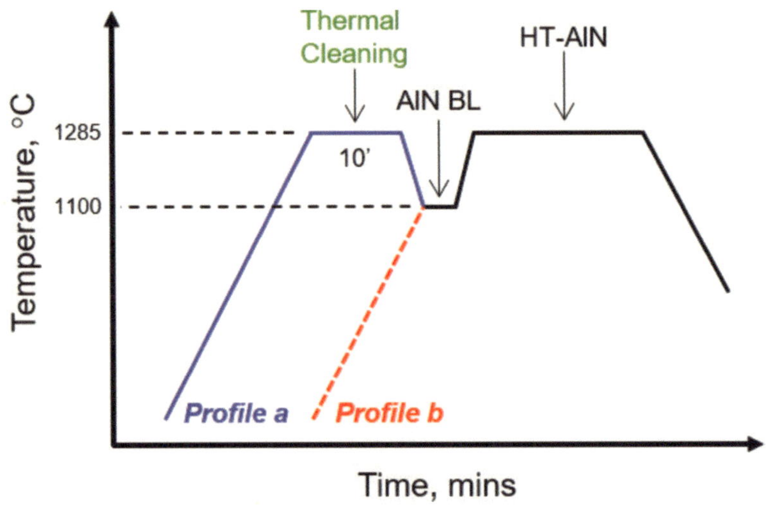

Figure 4. Temperature profiles for two-step growth of AlN on sapphire substrate. *Profile a* incorporates thermal cleaning, while *Profile b* incorporates no thermal cleaning.

4. Results and discussion

Let us then discuss the surface structure of the substrate after thermal cleaning under *Profile a* and before LT-AlN BL growth under *Profile b*. (For this case, the substrate was immediately cooled down, removed from the reactor, and analyzed.) As confirmed by AFM measurements, the thermal annealing under *Profile a* produced a parallel step-and-terrace surface structure with monolayer steps (**Figure 5(a)**). The average terrace-width and step height are ~125 and ~0.21 nm, respectively, as estimated from AFM line scan profile. Hence, the substrate's offcut angle was estimated to be ~0.11°, which is comparable to the expected off-cut angle of 0.15°. Moreover, because the ML step height corresponds to 1.299 nm $c_o/6$ = 0.22 nm, where c_o is the unit cell of sapphire having six ML steps of oxygen layers along the *c* axis [27], this confirms that the steps are correlated with the periodicity of oxygen and the interaction between oxygen atoms of successive layers is stronger than Al atoms of successive layers [26, 28].

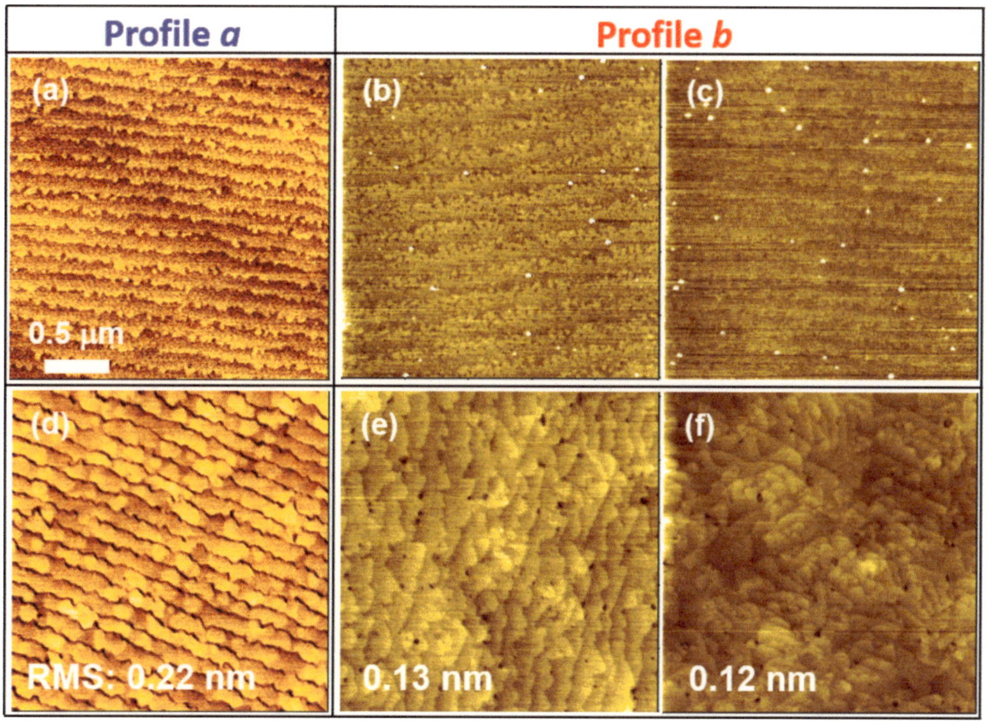

Figure 5. AFM surface morphologies of sapphire substrate after (a) thermal cleaning under *Profile a*, and before LT-AlN BL growth at (b) 1100°C and (c) 800°C under *Profile b*. The corresponding (d) HT-AlN growth for thermally cleaned substrate and under *Profile b* using (e) 1100°C and (f) 800°C LT-AlN BL T_g.

On the other hand, for sapphire substrate under *Profile b*, the formation of surface steps are undefined at BL T_g= 800°C (**Figure 5(c)**). However, two monolayer steps can be defined when the BL T_g is increased to 1100°C (**Figure 5(b)**). This indicates that with increasing BL T_g, the surface structure changes from "rough" to smooth having two ML steps. Therefore, different surface structures of sapphire substrate are formed depending on these two experimental conditions. It is therefore interesting to find out its effect on the structural quality of the subsequently grown AlN epilayer.

The corresponding surface morphologies for AlN grown under *Profile a* (**Figure 5(d)**) and *Profile b* at LT-AlN BL T_g=1100°C (**Figure 5(e)**) and 800°C (**Figure 5(f)**) are also shown in **Figure 5**. It can be seen that both profiles show an AlN with atomically smooth surfaces, as evidenced by their root-mean-square (RMS) roughness values. Moreover, a step-and-terrace surface morphology structure which replicates that of the substrate is exhibited for AlN grown under *Profile a*, while a meandering surface morphology is observed for AlN grown under *Profile b* using LT-AlN BL T_g=800°C (**Figure 5(f)**). Furthermore, these meandering steps begin to align along a certain direction as LT-AlN BL T_g is increased to 1100°C (**Figure 5(e)**). Clearly, these differences in AlN morphologies are most likely influenced by the surface structure of the substrate prior to AlN growth.

Figure 6 shows the asymmetric (10-12) ϕ-scan of AlN epilayers grown under *Profile a* (**Figure 6(a)**) and *Profile b* using LT-AlN BL T_g=1100°C (**Figure 6(b)**). The (11-23) ϕ-scan of the substrate is also shown. It is well known that the in-plane epitaxial relationship between AlN and sapphire is AlN<10-10>||α-Al_2O_3<11-20> [8]. Moreover, a closer look at AlN (10-12) diffraction shows two peaks under *Profile a* (**Figure 6(a)**) with separation $\Delta\theta$ ~3.72°, while only a single diffraction peak is observed under *Profile b*. This indicates the presence of a special grain boundary under *Profile a*, where the two AlN grains have a particular in-plane misorientation relationship, while it is successfully suppressed under *Profile b*. Furthermore, because sapphire (11-23) diffraction does not exhibit two peaks, the special grain boundary is confirmed to exist only in AlN epilayer.

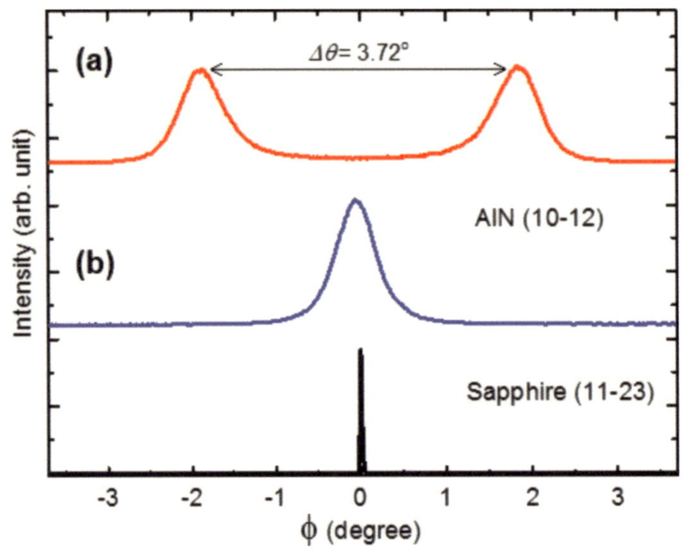

Figure 6. XRD ϕ-scans of AlN grown sapphire substrate under (a) *Profile a* and (b) *Profile b* using LT-AlN BL T_g = 1100°C.

To further confirm the existence of the special grain boundary in AlN under *Profile a*, a plan-view bright-field TEM micrograph and the corresponding selected area electron diffraction pattern (SAEDP) were taken under [0001] zone axis, as shown respectively in **Figures 7(a)** and **(b)**. A periodic bright and dark contrast of two AlN grains is observed. The AlN grain width is found identical to the step width of thermally cleaned sapphire substrate, implying that the origin of the grain boundary is related to the substrate's surface structure. As the AlN is grown onto sapphire substrate with either *A* or *B* oxygen stacking, the characteristic of that surface is also carried into AlN [26]. And as supported by XRD results and because only a slight misorientation relative to one another exists between these two AlN grains, this grain boundary is confirmed to be a small-angle grain boundary [25, 26]. Furthermore, due to the arrays of edge dislocations that exist at the boundary, this type of special boundary is called pure low-angle tilt SAGB. The spacing *D* between adjacent edge dislocation array can also be estimated using the formula $D = \mathbf{b}/\sin\theta \approx \mathbf{b}/\sin\Delta\theta$, where **b** is the in-plane burger's vector

(= 0.3112 nm) and $\Delta\theta$ is the misorientation angle ($\Delta\theta$ ~ 3.72°) obtained from XRD measurement. Hence, the spacing between edge dislocations is estimated to be ~4.75 nm. The SAEDP also supports the observation of SAGB, as seen from the double diffraction spots (denoted by arrow marks in **Figure 7(b)**). On the other hand, no special grain boundary is observed for AlN grown under *Profile b* (not shown). This suggests that the buffer layer technique is effective for suppressing the SAGB.

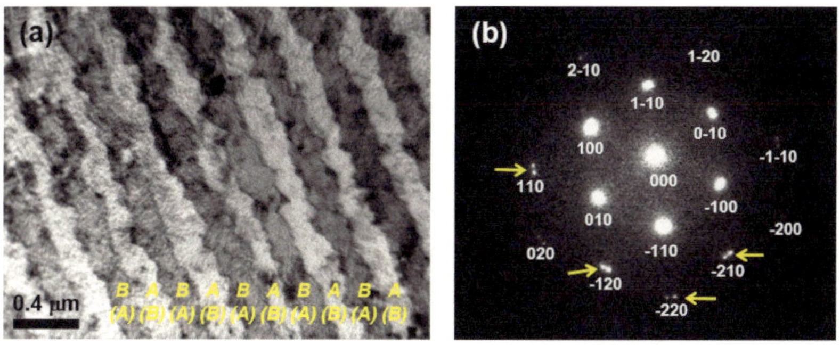

Figure 7. Plan-view TEM bright-field image of AlN grown under (a) *Profile a*. (b) Corresponding selected-area electron diffraction pattern of AlN in (a).

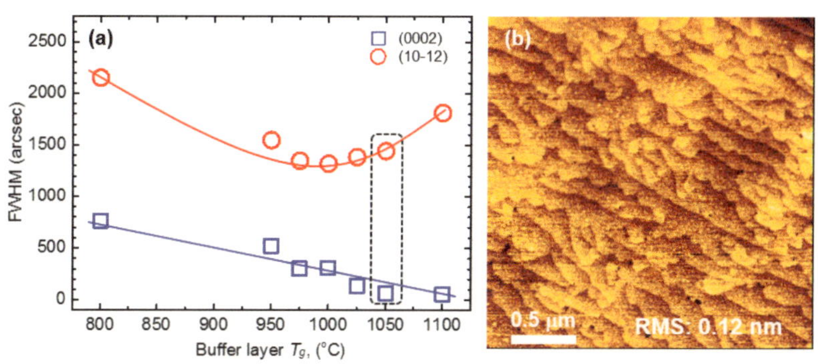

Figure 8. (a) LT-AlN BL T_g as a function of symmetric (0002) and asymmetric (10-12) ω-scans. (b) AFM surface morphology of AlN grown at optimum LT-AlN BL T_g (1050°C).

Then the LT-AlN BL T_g was optimized to improve the mosaicity of the epitaxial film using the current growth condition. The result of the XRD symmetric (0002) and asymmetric (10-12) ω-scans are shown in **Figure 8**. While the XRD linewidth of the tilt component corresponding to the symmetric (0002) becomes narrow with increasing temperature, the twist component corresponding to the asymmetric (10-12) initially becomes narrow and then broadens with increasing temperature. Hence, the narrowest linewidth for each component does not coincide with each other, and the reason is still unknown at this time. But by balancing these components, it can be deduced that the optimum LT BL temperature is T_g ~1050°C, where the XRD linewidths are ~66 and ~1443 arcsec, respectively, for (0002) and

(10-12) ω-scans. The symmetric component is comparable to that grown under *Profile a* (FWHM is ~64 arcsec), suggesting that both have highly-oriented films along the *c*-axis growth direction. On the other hand, the asymmetric (10-12) linewidth is wider than that grown under *profile a* (FWHM is 1145 arcsec). **Figure 8(b)** shows the AFM surface morphology of AlN grown under the optimum LT-AlN BL T_g of 1050°C. Although nanopits have been reduced compared with other BL T_g (**Figures 5(e), (f)**), it is believed that the quality of the film can be further improved especially the twist mosaicity upon optimizing the buffer layer thickness or other growth parameters.

To demonstrate the effect of eliminating the SAGB on the optical properties of AlN epilayer, we obtained the CL spectra for both profiles, as shown in **Figure 9**. The CL measurements were acquired at 93 K under 10 kV and 0.1 μA emission condition (spot size is ~1 μm). CL peaks are assigned to free and bound excitonic emissions, including the LO phonon replicas, as shown in the inset figure [14]. Moreover, the emission intensity of AlN under *Profile b* using the optimum LT-AlN BL T_g of 1050°C is approximately more than two times higher than that under *Profile a*. This result is attributed to the higher probability of radiative recombination of electron and hole pairs due to the elimination small-angle grain boundary. The differences in the CL peak position also suggest their different residual strain, where the tensile-strained AlN under *Profile b* has smaller bandgap energy (~6.011 eV) than that of an almost relaxed AlN (~6.031 eV) under *Profile a*. The slow relaxation experienced by AlN under *Profile b* is most likely due to the reduced generation of dislocations upon the suppression of SAGB. As SAGB is a type of an edge dislocation, eliminating it is expected to enhance the optical properties of AlN.

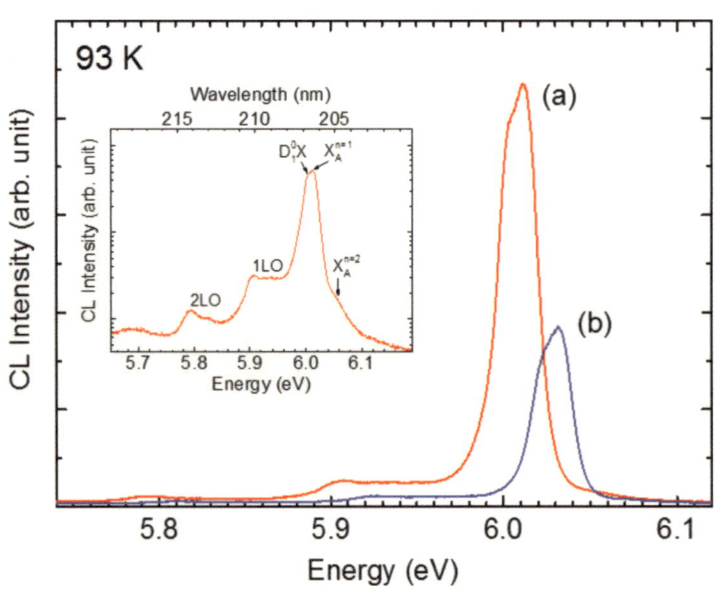

Figure 9. CL spectra of AlN grown under (a) *Profile a* and (b) *Profile b* using optimum LT-AlN BL T_g = 1050°C.

Hence, the origin of SAGB can be ascribed to the surface structure of sapphire substrate. It was shown that as long as the surface has ML steps, even the introduction of LT-AlN BL would be ineffective in eliminating the SAGB. Hence, there is a maximum allowable BL T_g at which the substrate's surface does not transform into monolayer steps yet. In this study, we consider that the maximum BL T_g would be around 1100°C. The result also implies that there is a proper timing to introduce the LT-AlN BL [21]. This tendency is also observed in direct HT-AlN growth using different growth methods [16, 21, 25], suggesting the wide observation of this phenomenon. It is noteworthy that even during temperature ramp-up, because the optimal T_g for HT-AlN growth is in the vicinity around $T_g = 1200$°C, it is likely that the surface could also change to periodic (ML) steps, as pointed out [25]. This is also supported in the present study based on the evolution of surface structure in both profiles. Therefore, the LT-AlN BL would be necessary in order to prevent the surface from transforming into structures with monolayer steps. Conversely, with the application of LT-AlN BL, the two ML steps observed using LT-AlN BL $T_g = 1100$°C is kept, hence, preserving their heteroepitaxial relationship (as expected, 2 ML steps do not form SAGB). For BL $T_g = 800$°C, it is believed that the "weak" heteroepitaxial relationship due to the undefined (rough) surface plays a role in circumventing the SAGB.

On the other hand, the SAGB can also be prevented even when using thermally cleaned substrate. In fact, a lot of AlN growth optimizations have been performed under such condition and yet, there was no observation of SAGB or hardly any mention if at all. The reason could be due to the weakening of the epitaxial relationship between the sapphire and AlN after introducing some extrinsic factors such as nitridation before LT-AlN BL or AlN seeding layer [11, 14, 26]. This could be the reason why the growth interruption by V/III ratio resulted in not only dislocation bending or coalescence but also the elimination of domain structure [18, 24]. Moreover, the insertion of an intermediate layer after LT-AlN BL could have also eliminated the rotation domain [30]. In addition, the no observation of rotation domain on thermally cleaned substrate would also depend on thermal cleaning temperature or temperature ramp-up. In their study, substrate surface after thermal cleaning at 1145°C prior to LT-AlN BL growth could have 2 ML step structure, thus preventing the SAGB [23]. Also, the initial supply of TMA (TMA preflow) during the pulse growth could have replaced the oxygen-terminated surface into Al-terminated one [25]. As mentioned earlier, the sapphire's surface with Al termination is expected to have no rotation domain.

Finally, **Table 1** shows the comparison between several techniques for eliminating the SAGB performed on as-received and nitrided sapphire substrates. Note that techniques included in the table are only those that directly discussed the rotation domain. Although a high-quality AlN could be obtained from nitrided sapphire substrate, residual SAGB still exists, albeit coalesced with other domain upon nitrogen radical or annealing treatment [19, 20]. Moreover, the step bunching of sapphire substrate into $n \times 2$ ML, where n is an integer, would require a critical off-cut angle during *ex situ* surface treatment [26]. On the other hand, the LT-AlN BL technique would eliminate these additional processes owing to its in situ treatment. It is noteworthy that although rotation domain phenomenon was not observed in previous studies using LT-AlN BL, we believe that whether intentionally or unintentionally, the elimination of rotation domain is one of its underlying purposes. For example, Xi et al. adopted this approach, and a high-quality and atomically smooth AlN epilayer was obtained [22]. The same is true in the work of Zhang et al.; however, they introduced the PALE approach to enhance the migration of Al adatoms during the HT-AlN growth [15]. But Hu et al. modified this method by performing nitridation pretreatment under H_2/NH_3 ambient prior to LT-AlN BL growth. However, they ascribed the elimination of rotation domain to the different strain relaxation mechanisms induced by lattice mismatch [21]. In addition, nitridation pretreatment may induce rough AlN surface. Therefore, LT-AlN BL is a promising technique for eliminating the rotation domain. It is noteworthy that LT BL has been used as a standard technique for the growth of GaN, resulting in crack-free and high-quality epilayer [31, 32]. However, to the best of our knowledge, this is the maiden report that clarifies the role of LT-AlN BL for obtaining high-quality AlN without SAGB.

Sapphire substrate	Pre/post-treatment technique	Suppression of SAGB	Smooth surface	In situ treatment
As-received	LT AlN BL without thermal cleaning (present technique)	+	+	+
As-received	TMA preflow [25]	+	+	+
	NH_3 preflow [25]	-	+	+
As-received	LT AlN BL with thermal cleaning [21]	-	+/-	+
	Nitridation and LT AlN BL [21]	+	+/-	+
As-received	Thermal cleaning [16, 26]	-	+	+
As-received	Annealing in air [26]	+	+	-
Nitrided	Annealing [20]	-	-	+
Nitrided	Nitrogen radical treatment [19]	-	-	+

Table 1. Comparison among several techniques reported for eliminating SAGB in AlN, where "+" sign shows the satisfied property, and the "-" sign shows the unsatisfied property.

5. Summary

In summary, the surface structure of sapphire substrate is found to influence the formation of small-angle grain boundary in subsequently grown AlN epilayer. The small-angle grain boundary is formed when the surface of the sapphire (0001) substrate is terminated by *ABAB* oxygen stacking with monolayer steps, which is formed during high-temperature thermal cleaning. To circumvent the small-angle grain boundary, the LT-AlN BL is introduced in order to circumvent the substrate structure from having monolayer steps. Rather, the substrate surface produced either rough structure at low BL T_g or defined two-ML-step structure as LT-AlN BL T_g is increased. CL measurement showed an increased emission from the AlN without SAGB. Thus, the LT-AlN BL technique would be effective in eliminating the SAGB, thereby obtaining high-quality AlN epilayer with improved optical and electrical properties.

Acknowledgements

The authors acknowledge Dr. M. Sumiya, Dr. K. Watanabe, and Dr. N. Ishida for their helpful discussions and assistance throughout this work.

Author details

Ryan G. Banal*, Masataka Imura* and Yasuo Koide*

*Address all correspondence to: BANAL.Ryan@nims.go.jp, IMURA.Masataka@nims.go.jp and KOIDE.Yasuo@nims.go.jp

Wide Bandgap Materials Group, National Institute for Materials Science (NIMS), Namiki, Tsukuba, Ibaraki, Japan

References

[1] Grandusky J.R., Smart J.A., Mendrick M.C., Schowalter L.J., Chen K.X., Schubert E.F. Pseudomorphic growth of thick n-type AlxGa1-xN layers on low-defect-density bulk AlN substrates for UV LED applications. Journal of Crystal Growth. 2009;311(10):2864–2866. DOI: 10.1016/j.jcrysgro.2009.01.101

[2] Rice A., Collazo R., Tweedie J., Dalmau R., Mita S., Xie J., et al. Surface preparation and homoepitaxial deposition of AlN on (0001)-oriented AlN substrates by metalorganic chemical vapor deposition. Journal of Applied Physics. 2010;108:043510:1–8. DOI: 10.1063/1.3467522

[3] Sugahara T., Sato H., Hao M., Naoi Y., Kurai S., Tottori S., et al. Direct evidence that dislocations are non-radiative recombination centers in GaN. Japanese Journal of Applied Physics. 1998;37(Part 2, Number 4A):L398–L400. DOI: 10.1143/JJAP.37.L398

[4] Khan A., Balakrishnan K., Katona T. Ultraviolet light-emitting diodes based on group three nitrides. Nature Photonics. 2008;2:77–84. DOI: 10.1038/nphoton.2007.293

[5] Shatalov M., Sun W., Jain R., Lunev A., Hu X., Dobrinsky A., et al. High power AlGaN ultraviolet light emitters. Semiconductor Science and Technology. 2014;29(8):084007. DOI: 10.1088/0268-1242/29/8/084007

[6] Hirayama H., Maeda N., Fujikawa S., Toyoda S., Kamata N. Recent progress and future prospects of AlGaN-based high-efficiency deep-ultraviolet light-emitting diodes. Japanese Journal of Applied Physics. 2014;53(10):100209:1–10. DOI: 10.7567/JJAP.53.100209

[7] Li X.-H., Kao T.-T., Satter Md.M., Wei Y.O., Wang S., Xie H., et al. Demonstration of transverse-magnetic deep-ultraviolet stimulated emission from AlGaN multiple-quantum-well lasers grown on a sapphire substrate. Applied Physics Letters. 2015;106:04115:1–4. DOI: 10.1063/1.4906590

[8] Vispute R.D., Wu H., Narayan J. High quality epitaxial aluminum nitride layers on sapphire by pulsed laser deposition. Applied Physics Letters. 1995;67:1549–1551. DOI: 10.1063/1.114489

[9] Hiroki M., Kobayashi N. Flat Surfaces and interfaces in AlN/GaN heterostructures and superlattices grown by flow-rate modulation epitaxy. Japanese Journal of Applied Physics. 2003;42(Part 1, Number 4B):2305–2308. DOI: 10.1143/JJAP.42.2305

[10] Adivarahan V., Sun W.H., Chitnis A., Shatalov M., Wu S., Maruska H.P., et al. 250 nm AlGaN light-emitting diodes. Applied Physics Letters. 2004;85:2175–2177. DOI: 10.1063/1.1796525

[11] Takeuchi M., Shimizu H., Kajitani R., Kawasaki K., Kumagai Y., Koukitu A., et al. Improvement of crystalline quality of N-polar AlN layers on c-plane sapphire by low-pressure flow-modulated MOCVD. Journal of Crystal Growth. 2007;298:336–340. DOI: 10.1016/j.jcrysgro.2006.10.140

[12] Hirayama H., Yatabe T., Noguchi N., Ohashi T., Kamata N. 231 – 261 nm AlGaN deep-ultraviolet light-emitting diodes fabricated on AlN multilayer buffers grown by ammonia pulse-flow method on sapphire. Applied Physics Letters. 2007;91:071901:1–3. DOI: 10.1063/1.2770662

[13] Zhang J.P., Khan M.A., Sun W.H., Wang H.M., Chen C.Q., Fareed Q., et al. Pulsed atomic-layer epitaxy of ultrahigh-quality AlxGa1-xN structures for deep ultraviolet emissions below 230 nm. Applied Physics Letters. 2002;81:4392–4394. DOI: 10.1063/1.1528726

[14] Takeuchi M., Ooishi S., Ohtsuka T., Maegawa T., Koyama T., Chichibu S.F., et al. Improvement of Al-Polar AlN layer quality by three-stage flow-modulation metalorganic chemical vapor deposition. Applied Physics Express. 2008;1(Number 2): 021102:1–021102:3. DOI: 10.1143/APEX.1.021102

[15] Zhang J.P., Wang H.M., Sun W.H., Adivarahan V., Wu S., Chitnis A., et al. High-quality AlGaN layers over pulsed atomic-layer epitaxially grown AlN templates for deep ultraviolet light-emitting diodes. Journal of Electronic Materials. 2003;32(5):364–370. DOI: 10.1007/s11664-003-0159-2

[16] Banal R.G., Funato M., Kawakami Y. Initial nucleation of AlN grown directly on sapphire substrates by metal-organic vapor phase epitaxy. Applied Physics Letters. 2008;92:241905:1–241905:3. DOI: 10.1063/1.2937445

[17] Ohba Y., Yoshida H., Sato R. Growth of High-Quality AlN, GaN and AlGaN with atomically smooth surfaces on sapphire substrates. Japanese Journal of Applied Physics. 1997;36(Part 2, Number 12A):L1565–L1567. DOI: 10.1143/JJAP.36.L1565

[18] Imura M., Nakano K., Fujimoto N., Okada N., Balakrishnan K., Iwaya M., et al. High-temperature metal-organic vapor phase epitaxial growth of AlN on sapphire by multi transition growth mode method varying V/III ratio. Japanese Journal of Applied Physics. 2006;45(Part 1, Number 11):8639–8643. DOI: 10.1143/JJAP.45.8639

[19] Ueno K., Ohta J., Fujioka H., Fukuyama H. Characteristics of AlN films grown on thermally-nitrided sapphire substrates. Applied Physics Express. 2010;4(Number 1): 015501:1–015501:3. DOI: 10.1143/APEX.4.015501

[20] Adachi M., Sugiyama M., Tanaka A., Fukuyama H. Elimination of rotational domain in AlN layers grown from Ga–Al flux and effects of growth temperature on the layers. Materials Transactions. 2012;53(Number 7):1295–1300. DOI: 10.2320/matertrans.MBW201112

[21] Wang H., Xiong H., Wu Z.-H., Yu C.-H., Tian Y., Dai J.-N., et al. Occurrence and elimination of in-plane misoriented crystals in AlN epilayers on sapphire via pre-treatment control. Chinese Physics B. 2013;23(Number 2):028101:1–028101:5. DOI: 10.1088/1674-1056/23/2/028101

[22] Xi Y.A., Chen K.X., Mont F., Kim J.K., Wetzel C., Schubert E.F., et al. Very high quality AlN grown on (0001) sapphire by metal-organic vapor phase epitaxy. Applied Physics Letters. 2006;89:103106:1–103106:3. DOI: 10.1063/1.2345256

[23] Lai M.-J., Chang L.-B., Yuan T.-T., Lin R.-M. Improvement of crystal quality of AlN grown on sapphire substrate by MOCVD. Crystal Research and Technology. 2010;45(7): 703–706. DOI: 10.1002/crat.201000063

[24] Peng M.-Z., Guo L.-W., Zhang J., Yu N.-S., Zhu X.-L., Yan J.-F., et al. Three-step growth optimization of AlN epilayers by MOCVD. Chinese Physics Letters. 2008;25(Number 6):2265–2268. DOI: 10.1088/0256-307X/25/6/094

[25] Kawaguchi K., Kuramata A. Defect structures of AlN on sapphire (0001) grown by metalorganic vapor-phase epitaxy with different preflow sources. Japanese Journal of Applied Physics. 2005;44(Part 2, Numbers 46–49):L1400–L1402. DOI: 10.1143/JJAP.44.L1400

[26] Hayashi Y., Banal R.G., Funato M., Kawakami Y. Heteroepitaxy between wurtzite and corundum materials. Journal of Applied Physics. 2013;113:183523:1–183523:8. DOI: 10.1063/1.4804328

[27] Chiang Y.-M., Birnie III D., Kingery W.D. Physical Ceramics: Principles for Ceramic Science and Engineering. Canada: Wiley & Sons, Inc.; 1997. 540 p.

[28] Van L.P., Kurnosikov O., Cousty J. Evolution of steps on vicinal (0001) surfaces of α-alumina. Surface Science. 1998;411(3):263–271. DOI: 10.1016/S0039-6028(98)00329-X

[29] Berghezan A., Fourdeaux A., Amelinckx S. The structure and properties of grain boundaries. Acta Metallurgica. 1961;9(5):464–490. DOI: 10.1016/0001-6160(61)90142-0

[30] Cai B., Nakarmi M.L. TEM analysis of microstructures of AlN/sapphire grown by MOCVD. In: Gwo S., Ager J.W., Ren F., Ambacher O., Schowalter L., editors. MRS Proceedings; November 30–December 4, 2009; Boston, Massachusetts. Materials Research Society; 2010. pp. 1202-I05:01–1202-I05:05. DOI: 10.1557/PROC-1202-I05-01

[31] Amano H., Sawaki N., Akasaki I., Toyoda Y. Metalorganic vapor phase epitaxial growth of a high quality GaN film using an AlN buffer layer. Applied Physics Letters. 1986;48:353–355. DOI: 10.1063/1.96549

[32] Nakamura S. GaN growth using GaN buffer layer. Japanese Journal of Applied Physics. 1991;30(Part 2, Number 10A):L1705–L1707. DOI: 10.1143/JJAP.30.L1705

Chapter 4

Grain Boundary in Oxide Scale During High-Temperature Metal Processing

Xianglong Yu and Ji Zhou

Additional information is available at the end of the chapter

http://dx.doi.org/10.5772/66211

Abstract

Grain boundary in oxide scale has profound influences on the deformation behaviour and tribological properties of metal alloys at high temperature. This chapter introduces some recent progress to quantify microstructure and interface quality, providing examples of possible property variations. Some fundamental issues of oxidation mechanism have been given, consisting of crystal structures of iron oxides and oxidation of steel alloys. Two main things are addressed: One is what the characters of grain boundaries are developed in the oxide scale, which is associated with grain shape and size, microtexture, and special grain boundaries such as coincident site lattice (CSL) boundaries. Another is the role of grain boundaries played during metal processing, including initial oxidation via grain boundary diffusion, stress and deformation processing, and tribological properties of oxide scale at metal processing. Finally, a more extensive effort was also made to summarise the experimental techniques used to investigate oxide scale.

Keywords: grain boundary, oxide scale, oxidation, steel, metal processing

1. Introduction

Metallic oxides are inevitable to generate on the surface of steel products during metal forming processes at high temperature. If its thickness is larger than nanometre size, the oxide layer can be termed as *oxide scale* in industrial manufacturing. The formation of oxide scale or the oxidation of industrial materials depends, not only on the properties of their component crystals, but also on those of the boundaries between those crystals, in particular the structure and chemical composition of the boundaries.

Grain boundaries have distinct properties relative to bulk material in terms of atomic coordination, reactivity and diffusion rates. Phase contact creates interfaces that represent changes in composition with disrupted atomic bonding. Grain boundary refers to the interface zone between grains of the same phase, while interfaces are boundaries between dissimilar phases. It represents the narrow zone where atomic bonding is disrupted by misalignment of the crystalline grains. This disrupted bonding at the grain boundary is about 5–10 atoms across. Grain boundaries have nanoscale spatial dimensions, which can generate substantial resistance to ionic transport due to dopant (or impurity) segregation. This diffusion provides active paths for atomic motion, particularly at high temperature, during the diffusion-controlled oxidation or corrosion. Thus, composite properties are sensitive to the interface structure and chemistry to large potential variations.

This crystallographic structure of a metal alloy is one of the important parameters in determining the oxidation or corrosion behaviour. The characters of grain boundaries in oxide layers formed on substrates influence adhesion and friction behaviour, surface fracture and wear during high temperature steel processing. However, the effect of grain characters on the oxidation behaviour is not fully understood yet. There are still many challenges, one of which is how to engineer grain boundaries to optimise the oxidation resistance of these materials. For this reason, detailed understanding of the processing-structure-property relationships that focus on grain boundaries and interfaces is critical to advanced manufacturing of metals. Furthermore, it is necessary to modify the grain boundary characteristics of this alloy which affect its oxidation resistance.

In this chapter, an attempt has been made to explore the role and behaviour of grain boundaries in the oxide scale formed on the steel surface during metal processing. In doing so, two things we need to consider for such high-temperature plastic deformation are diffusion mechanism at grain boundaries and resulting boundary migration in the growth of grains.

2. High-temperature oxidation

This section is devoted to the fundamental issues of oxidation mechanism that should be defined and summarised before specific problems are confronted. During metal processing at elevated temperatures, oxidation occurs inevitably on the surface of products. In the case of the pure iron, the oxide scale formed on is a complex mixture of three iron oxide phases: hematite (Fe_2O_3), magnetite (Fe_3O_4) and wustite ($Fe_{1-x}O$, x = 0.84–0.95) [1]. This is because iron has divalent and trivalent ions (Fe^{2+} and Fe^{3+}). The complete oxidation of iron can be divided into three main steps, where iron oxidises to the lowest valence ion Fe^{2+} and forms the first sub-layer of wustite ($Fe_{1-x}O$) next to the metal. Then, some of Fe^{2+} ions oxidise further to Fe^{3+} and contain both valence iron ions as the intermediate sub-layer of magnetite (Fe_3O_4). Under conditions of sufficient oxygen, the outer sub-layer of hematite (Fe_2O_3) only consists of the highest valence iron ion Fe^{3+}. This is the case above 570°C (the eutectoid point of the Fe-O system) in the diffusion-controlled growth of multilayered scales on pure iron. Below 570°C, the wustite phase is unstable, and the oxidation of iron directly results in magnetite. In steel, various Fe-C alloys, their oxidation at high temperature can be more complex than pure iron, in particular the segregation of different element at grain boundaries.

2.1. Iron oxides

In view of the iron cations can exist in the two valence states, iron oxides can have different crystal structures with different Fe/O ratios. These phases include wustite ($Fe_{1-x}O$), magnetite (Fe_3O_4) and hematite (α-Fe_2O_3) [2].

Wustite has a defective halite structure, with anion sites occupied by O^{2-} and most cation sites occupied by divalent Fe^{2+} ions. Cubic close-packed (CCP) array of O^{2-} stacked along the [111] direction. Most of the iron is octahedral with a small proportion of Fe^{3+} on the vacant tetrahedral sites (**Figure 1a**). A cation-deficient phase written as $Fe_{1\_x}O$ (with 1_x ranging from 0.83 to 0.95) exists at 0.1 MPa pressure and temperatures higher than 570°C. Wustite is a p-type (p = positive carrier) semi-conducting oxide with a high concentration of lattice defects. These high cation vacancies result in a high mobility of cations and electron via metal vacancies and electron holes.

Magnetite (Fe_3O_4) has an inverse spinel structure containing both divalent and trivalent iron ions. The distribution of its cation is written as (Fe^{3+}) [$Fe^{3+}Fe^{2+}$] O_4, where the parentheses denote the tetrahedral sites and the square brackets denote the octahedral sites. In this case, the ferric ion Fe^{3+} relinquishes half of the octahedral sites to the ferrous species Fe^{2+}, that is, with $8Fe^{3+}$ ions located in tetrahedral sites plus ($8Fe^{3+}$ and $8Fe^{2+}$) ions distributed into octahedral sites per unit cell. The structure consists of octahedral and mixed tetrahedral/octahedral layers stacked along the [111] direction. **Figure 1b** shows the sequence of Fe- and O-layers and the section of this structure with three octahedral and two tetrahedral. Magnetite with an excess of oxygen also exists, but this excess is much smaller than that with wustite, and the corresponding concentration of defects is also less.

The crystal system of hematite (α-Fe_2O_3) is a rhombohedral structure (**Figure 1c**) with a low concentration of structural defects. Hexagonal close-packed (HCP) arrays of oxygen ions are stacked at the [001] direction. The O-O distances along the shared face of an octahedron are shorter (0.2669 nm) than the distance along the unshared edge (0.3035 nm), and hence, the octahedron is distorted trigonally. The shared Fe-O_3-Fe triplet structure influences the

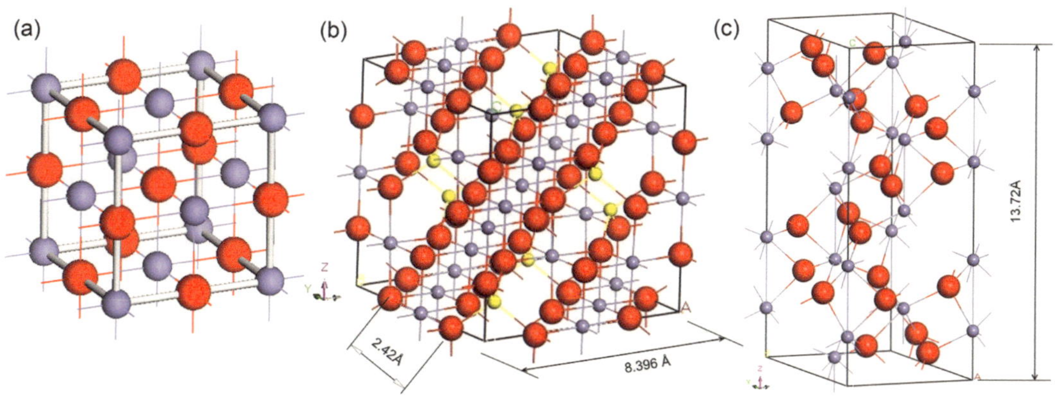

Figure 1. Crystal structure of (a) wustite (FeO), (b) magnetite (Fe_3O_4) and (c) hematite (α-Fe_2O_3).

magnetic properties of the oxide because hematite is an n-type (n = negative carrier) semiconducting oxide in which the diffusion of anions is dominant.

In a word, these iron oxides have the different crystal symmetries representing by a different space group: ferrite, *Im-3m*; wustite, *Fm-3m*; magnetite, *Fd-3m*; hematite, *R-3c*, and different lattice parameters: ferrite, wustite and magnetite in cubic symmetry with lattice parameters (a) of 0.287, 0.431 and 0.840 nm, respectively, and hematite in a trigonal structure with $a = 0.504$ nm and $b = 1.377$ nm [3].

2.2. Oxidation of steel alloys

Due to the presence of alloying elements and impurities in Fe-C steel alloys, the oxidation rates, phase development and morphologies of oxide scale are dramatically different from pure iron under various processing parameters. The reader can be referred to some published books [1], with regard to these four factors, alloying elements, oxidation kinetics, resulting oxides and their microstructure. More attention will be given here to the relationships between theories associated with grain boundaries of oxide scale.

The oxide scale formed on low carbon steels generally consists of a typical three-layered microstructure with a thin outer layer of hematite (Fe_2O_3), an intermediate layer of magnetite (Fe_3O_4) and an inner layer of wustite ($Fe_{1-x}O$, $1\_x = 0.84–0.95$) adjacent to the steel substrate [4]. This three-layered microstructure remains until a eutectoid point of the Fe-O phase equilibrium diagram is reached if oxygen is available at a temperature of 570°C. When the temperature drops below 570°C, the wustite phase becomes unstable and will decompose into eutectoid products of magnetite and ferrite. During the cooling from high temperature, the morphology and composition within oxide scale will be changed significantly. This variation mainly depends on heat treatments, atmosphere of the gas and chemistry of steels.

Analogous to the alloying strength in steel substrate, the alloying additions modified the oxidation quite dramatically. Normally, the amount of silicon and chromium aims to form a protective oxide scale, whereas small additions of nickel, copper, niobium, molybdenum and vanadium led to a greatly increased adherence of oxide scale. For instance, silicon generally enriches and forms an anchor-like morphology at the oxide scale/steel interface or grain boundaries of the oxides. Manganese is normally used as a solvent, while the transport of carbon is via defects such as pores rather than lattice or grain boundary diffusion. Because manganese has a stronger affinity for oxygen than iron does, manganese is normally spread sparingly over the entire oxide layer in the cross-sectional direction [3]. In addition to alloying elements, the atmosphere of gas as it interacted with oxides, especially water vapour, has made our understanding of the overall situation elusive [5], while the different behaviour of steel alloys in air-moisture mixtures has further complicated the set of observations.

The thickness or weight change in oxide scale with time is generally used to assess oxidation rates of the metallic alloys. In a linear or parabolic growth rate, the mixed para-linear kinetics is widely accepted to deal with the short-time growth of oxide scale during high-temperature metal processing. With the oxidation of steels below 727°C (the eutectoid point of the Fe-C system [4]), the oxidation kinetics is similar to pure iron due to without decarburisation at various atmospheres and follows approximately parabolic kinetics. In a case of oxidation in pure iron at 700–1200°C, the thickness ratio between 100:5:1 and 100:10:1 ($FeO:Fe_3O_4:Fe_2O_3$) can be obtained, whereas at 400 and 550°C without wustite, it can be 10:1 to 20:1 ($Fe_3O_4:Fe_2O_3$) [6]. The phase development during the high-temperature oxidation of steel alloys concentrated mostly on the evolution of wustite during isothermal holding because wustite will decompose into magnetite and ferrite below 570°C.

The oxide scale consists of most magnetite and hematite at room temperature. Three research directions can be: wustite formation above 700°C, magnetite and hematite below 570°C and wustite decomposition between two temperature ranges. This chapter here only focuses the morphologies of oxide scale at room temperature cooling from high-temperature processing. It is noted that this scheme, using microstructures at room temperature to deduce what happened at high temperature, can thus far be subject to the current characterisation techniques. If one *in situ* technique is available to observe oxidation behaviour at high temperatures around 1000°C, especially during a long term, some current results could be improved significantly.

3. Characteristics of grain boundaries in oxide scale

This section covers the microstructure or morphologies, the crystallographic preferred orientation (i.e. texture) and characteristics of grain boundaries in the magnetite/hematite oxide scale.

3.1. Microstructure characterisation

Figure 2 shows the oxidised samples in the cross-sectional or thickness direction parallel to the direction of oxide growth and from the top surface. Electron backscattered diffraction (EBSD) phase mapping shown in **Figure 2a** and **b** indicates a columnar-shape microstructure between the outer granular grains and the globular inner layer [7]. The oxide scale is composed of a thin outer layer of hematite and the inner duplex magnetite layers. The outer layer is columnar in structure, whereas the inner layer is much finer grained and the grains are equiaxed. The grains of magnetite have granular shape with the grain size around 3 μm in the outer layer of oxide scale. In addition, hematite near the surface gradually penetrates into the cracks within the oxide scale.

Grain shape and size highly influence the oxidation of pure metals and their alloys at high temperature. Oxidised scale shows the rough microstructure with valleys around grain boundary (**Figure 2c**, **d** [8]). This indicates that the transport of cations along grain boundaries is the dominant mechanism for outer scale growth. The diffusion of metal ions can result in vacancies and cavity to facilitate the formation of local pores [8]. Therefore, the grain-refined metal substrate can enhance the grain boundary diffusion at high temperature.

3.2. Microtexture evolution of oxide scale

Crystallographic orientation refers to how the atomic planes in a volume of crystal or grain are positioned relative to a fixed reference [9]. These grains present the occurrence of certain

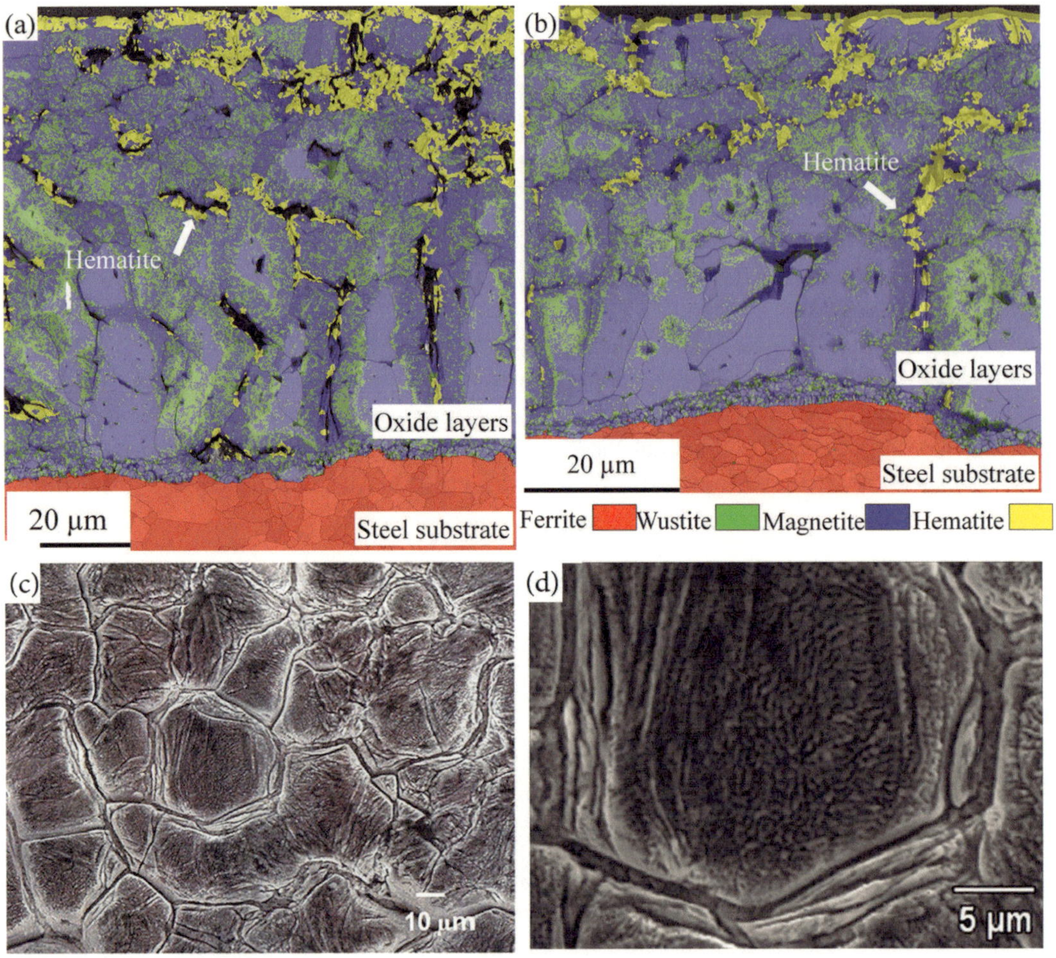

Figure 2. EBSD phase maps for wustite, hematite, magnetite and ferrite of the microalloyed steel hot rolled at 860°C with the thickness reductions and then cooling rates of (a) 10%, 10°C/s, and (b) 13%, 23°C/s [7], (c) SEM morphologies of top oxide scale formed over pure iron and (d) the magnification of one of valley surface [8].

orientations caused by heat treatments from melting and subsequently thermomechanical processing. This tendency is known as preferred orientation or texture. *Microtexture* is the conjoining of microstructure and texture [9], referring to the orientation statistics of a population of individual grains and their spatial location, that is, the orientation topography. To avoid ambiguity, a texture that reflects an average value obtained from many different grains is often called macrotexture [9]. Microtexture can generally be representing by pole figure, inverse pole figure or orientation distribution functions (ODF). Pole figure and inverse pole figure can be defined by the projection of orientation space through the crystal to specimen coordinate system or vice versa. ODF in the form of sections through the orientation space express the probability density function of orientations. This is a quantitative evaluation of the microtextures made by means of spherical harmonics method.

Analysis of the microtexture in the oxide phases and their orientation relationship is now being studied. A strong {001} texture may be found in wustite whatever the steel substrate [10], though this fibre texture also evolves in magnetite under low-temperature oxidation [6]. **Figure 3** shows texture development of magnetite and hematite in deformed oxide layers and their intensity distributions along associated fibres or texture components. Magnetite has a cubic structure, and its ODF sections are depicted using the $\phi_2 = 0°$ and 45° (**Figure 3b**) in terms of the Bunge system. In contrast, for hexagonal hematite, the ODF sections with $\phi_2 = 0°$ and 30° (**Figure 3d**) are used [9]. θ fibre develops in magnetite superimpose on $\phi_2 = 0°$ section at $\Phi = 0°$ with the rotations of <100>//ND [6, 11]. **Figure 3a** shows the intensity distribution of the θ fibre in magnetite on the samples subjected to various deformation conditions. <10-10> fibre component in α-Fe_2O_3 (**Figure 3c**) lies on $\phi_2 = 30°$ section and corresponds to orientations along $\Phi = 90°$.

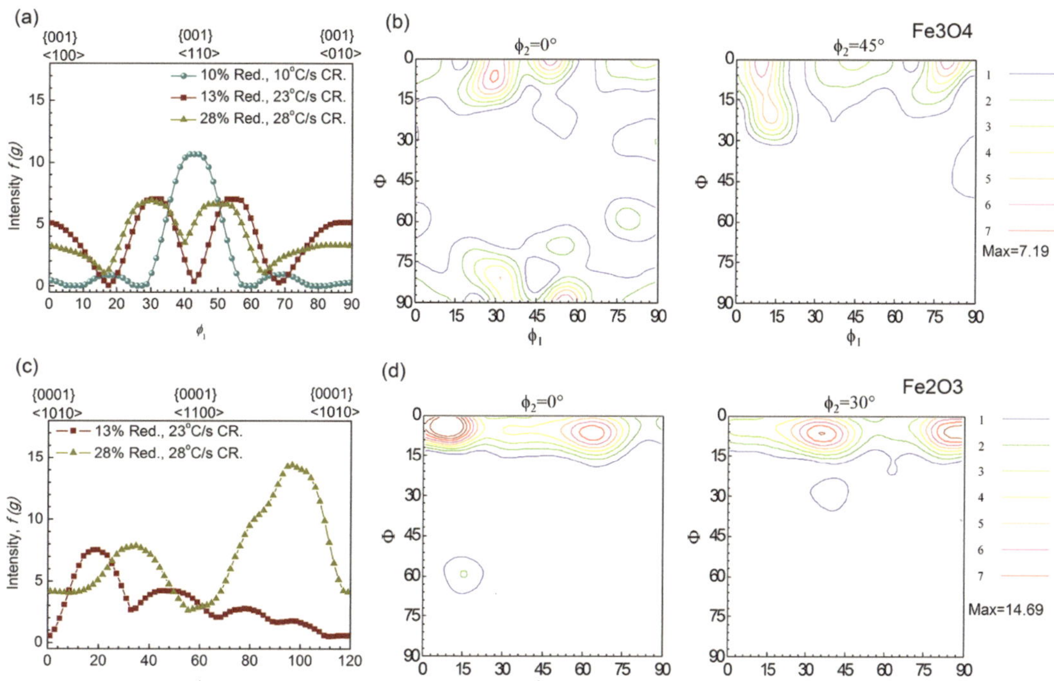

Figure 3. Development of texture intensity $f(g)$ along the (a) θ fibre of magnetite, (c) <1010> fibre of hematite, of the samples with different thickness reductions (TRs) and cooling rates (CRs), and orientation distribution function (ODF) sections for (b) magnetite and (d) hematite at a TR of 28% and a CR of 28°C/s [12].

Study on orientation relationship of oxide phase is still less explored thus far. A cube-cube orientation relationship between wustite and magnetite may prevail in undeformed oxide scale possibly due to the defective structure of the wustite. The orientation of the magnetite and substrate was reported {110}Fe//{100}Fe_3O_4, <110>Fe//<100>Fe_3O_4 in the case of transforming by continuous cooling from 400°C. By contrast, the Fe/FeO orientation relationship was {100}

Fe//{110}FeO, <110>Fe//<110>FeO. For a very thin oxide scale, a Fe/FeO orientation relationship was {100}Fe//{100}FeO, <100>Fe//<110>FeO [13].

3.3. Character distribution of grain boundaries

Effects of grain orientation and grain boundary characters on the elevated temperature oxidation behaviour demonstrate the role of grain boundaries in enhancing high-temperature oxidation resistance of various polycrystalline steel alloys. Overall surface energy and tribological behaviour can also be enhanced by grain boundary engineering.

3.3.1. Grain boundaries

Grain boundaries can be classified geometrically in terms of the relative misorientation between the neighboured grains. This relative misorientation can be defined by misorientation axis and angle. For instance, $2° \leq \theta < 15°$ misorientations are defined as low-angle grain boundaries (LAGBs), whereas the high-angle grain boundaries (HAGBs) are $\geq 15°$. Certain specific combinations resulting in a coincidence site lattice (CSL), the degree of coincidence is represented by the reciprocal density of common lattice points, denoted as the Σ number. As such, special grain boundaries refer to the low $\sum$ ($\sum \leq 29$) CSL boundaries, even though there is no physical basis for this assumption [14].

In oxidation and corrosion, it is widely believed that HAGBs have undergone hot corrosion and substantial depletion/segregation of alloying elements through the entire cross section. Distribution of grain boundaries in surface layer of oxide scale reveals that the misorientation tends to be large near grain boundaries, particularly at the oxide-substrate interface, where the high fraction of small magnetite grains is accumulated.

3.3.2. Special grain boundaries

Low-energy CSL boundaries with higher mobility can enhance the resistance of cracking or oxidation [15, 16]. CSL boundaries with low Σ orientation ($\Sigma \leq 49$) display improved physical and chemical properties relative to general or high CSL boundaries ($\Sigma > 49$) [17]. Some studies [18] reported that the resistance to intergranular oxidation of Ni-Fe alloy increased upon increasing the fraction of special boundaries. The extent of oxidation of individual Σ boundaries in Ni-Fe alloys is based on morphological observations. It found that $\Sigma 3$, $\Sigma 11$ and $\Sigma 19$ were more resistant to oxidation than other Σ boundaries [19].

In a oxidised microalloyed low carbon steel [12], a high proportion of low-angle and low-ΣCSL boundaries, magnetite for 60°/<111> ($\Sigma 3$), and hematite for 57.42°/<1-210> ($\Sigma 13$b) and 84.78°/<0-110> ($\Sigma 19$c) can be found. Misorientation peaks occur in α-Fe_2O_3 for axes near <0001> in the angle range of 27°–63° and <102> in the angle range of 63°–83°. For α-Fe_2O_3, the relatively high densities correspond to 57.42°/<110> ($\Sigma 13$b) and 84.78°/<010> ($\Sigma 19$c).

Furthermore, CSL boundaries distributions in **Figure 4** [17] reveal that Fe_3O_4 carries a high proportion of $\Sigma 3$, $\Sigma 5$ and $\Sigma 7$, whereas α-Fe_2O_3 has a profound fraction of $\Sigma 7$, $\Sigma 13$b and $\Sigma 19$c. It is noted that coherent twins have been excluded from this analysis, which results in a significantly lower fraction of $\Sigma 3$ boundaries. In any case, it becomes clear that these low CSL grain

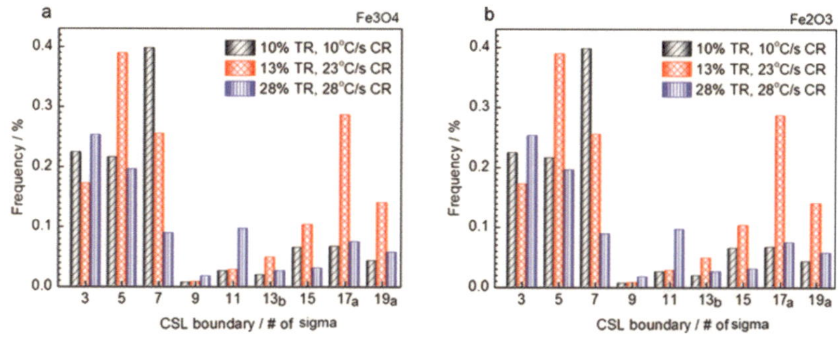

Figure 4. Histogram plots of CSL boundary distribution for (a) Fe_3O_4 and (b) α-Fe_2O_3, of the samples with different thickness reductions (TRs) and cooling rates (CRs) of oxide scale formed on a microalloyed low carbon steel [17].

boundary characteristics in Fe_3O_4 and α-Fe_2O_3 can be used to enhance crack resistance and further improve tribological properties of oxidised steels during high-temperature processing.

4. Roles of grain boundaries in oxide scale

The grain boundaries of either wustite, magnetite or hematite play a significant roles in the oxidation of metals and during their processing at high temperature. Three stages can be divided into: (i) diffusion-controlled oxidation of metal alloys; (ii) the plastic deformation mechanism near grain boundaries and resulting fracture of oxide scales; and (iii) tribological properties of oxide scale consisting various different grain boundaries during metal processing. All these above are this section will address.

4.1. Diffusion-controlled oxidation mechanism in grain boundaries

Grain characters, such as grain shape or grain boundary, highly influence the oxidation kinetics of pure metals and alloys. Grain boundary diffusion is more predominant in iron metal oxidation at high temperature. At low temperatures, the role of grain boundary diffusion as a main factor in comparison with other short circuits remains elusive.

To understand the role of grain boundary in diffusion-controlled oxidation, it is essential to detect which types of grain boundaries are involved. The EBSD/X-ray energy dispersive spectroscopy (EDS) map scanning can analyse the elemental distribution and correlate with the grain boundary character, and hence visualise the type of grain boundaries that are susceptible to hot corrosion or oxidation. **Figure 5** shows the image quality plus grain boundary map of the cross section of the hot corrosion alloy 617 [14] and EDS elemental distribution maps of various alloying elements. Preferential segregation/depletion of alloying elements occurred at grain boundaries: the segregation of Mo, S, Co and Ni at the random HAGBs along with a depletion of Cr after hot corrosion. The presence of S segregations also at intact interfaces and at oxide grain boundaries affects the oxide growth mechanism [20]. $\sum 3$ boundaries show few preferential enrichment/depletion of any alloying element, that is, indicating that these boundaries are resistant to hot corrosion.

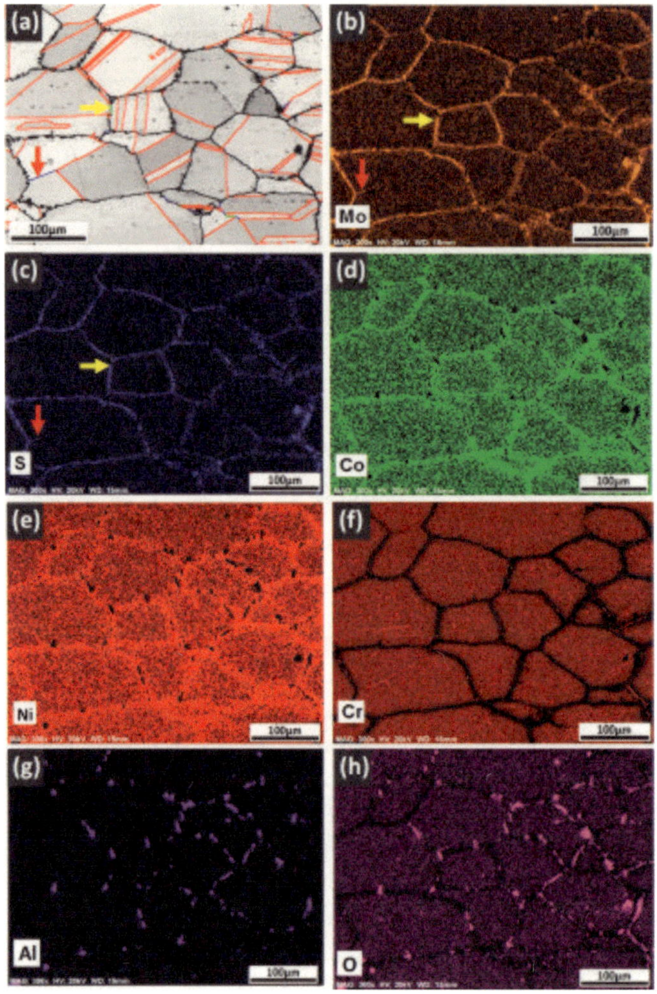

Figure 5. (a) Image quality plus grain boundary map (colour code: ∑3-red, ∑9-blue, ∑27-green, random HAGBs-black) and EDS elemental map showing distribution of (b) Mo, (c) S, (d) Co, (e) Ni, (f) Cr, (g) Al and (h) O across the cross section of the alloy 617 after hot corrosion testing [14].

Two dominant diffusion can occur at grain boundary or lattices [8]. Surface diffusion happens at lower temperatures compared to grain boundary diffusion, and volume diffusion is active only at very high temperatures. With a small grain size, the higher grain boundary area naturally increases grain boundary diffusion [21]. Grain boundary diffusion is more sensitive to grain size when compared to volume diffusion. In contrast to lattice diffusion, the control of elemental diffusion at the grain boundaries can be effective to have a thin and compact oxide scale on the Fe-Cr alloy surface [22]. This suggests that the grain boundary diffusion is confined at the initial of oxidation, while the oxide layer is relatively thin.

In a stainless steel of cyclic steam oxidation, the previous results [23] indicate that grain boundaries not only promote the chromium outward diffusion, but also provide the fast diffusion paths for the oxygen penetration. The grain boundaries promote the iron outward diffusion, accompanied with the fast growth of interfacial voids between two oxide layers. In a Ni-5Cr alloy, intergranular selective oxidation also accompanied by local chromium depletion and diffusion-induced grain boundary migration. Recently, coupled transmission electron microscope (TEM)/APT surface and grain boundary oxide compositions were identified, and Ni enrichment was observed around the oxides. The data provide novel information on the role of the minor impurities and the formation of early-stage oxides in 304 stainless steel [24]. However, copper diffusion along grain boundaries is not the main mechanism in this case. A high-resolution characterisation of the oxide–metal interface has shown the presence of a Fe-rich oxide, less dense than the original Cr-rich oxide [25]. It is reasonable that the Gibbs free energy reduction with Cu spinel solid solution formation in hematite at high oxygen partial pressure induces the bulk diffusion of Cu through hematite grains to the top surface of external oxide [26].

Various diffusion mechanisms can differ from types of grain boundaries in different oxidised substrates, for example, CSL special grain boundaries in ferritic stainless steel [18], whereas high-angle grain boundaries in Al_2O_3 [27]. Diffusion-controlled oxidation mechanism of the oxides thermally grown on the metal surface is similar to the pure oxides in bulk ceramics, ranging from a point effect mechanism to migration of disconnections, grain boundary ledge defects [27].

4.2. Deformation mechanism near grain boundaries

This section will discuss the internal stress state after diffusion-controlled oxidation of metal alloys and plastic deformation of oxide scales during metal processing. The occurrence of concomitant grain boundary sliding in the thermally grown oxides may be evidenced leading then to the corresponding microscopic strain. Local strain caused by the oxidation of magnetite to hematite can cause inter crystalline microcracks. These microcrackings can induce plastic deformation under differential contraction and to open diffusion paths inducing grain boundary diffusion.

Cracking propagation can roughly attribute to alloying elements segregation at grain boundaries. To delve, then which types of grain boundaries will occur these elements accumulation, and which types of alloying elements would be detrimental to crack propagation? For example, the Co oxide enriches at the boundaries of high stacking fault (SF)/low SF grains [28] and the Ni/Ti/Al-rich oxides at normal grain boundaries. But the enrichments of these elements have slightly influence on crack initiation and propagation in some Ni-based superalloy.

The mechanical stresses in the oxide scale play a significant role in its integrity. Generally, internal stresses are induced by the growth of oxides, thermal expansion mismatch and applied forces [6], some of which originate from many different causes. The formation and propagation of cracks generally occur along grain boundaries of oxide scale. The stress is the greatest at the tips of small cracks in the material, and consequently, the reaction proceeds at its greatest rate from these tips. To alleviate the propagation of cracks, low-angle and low-ΣCSL boundaries in microstructure can offer obstacles, because they minimise the solute effects and reduce the interaction between the interfaces and glissile dislocation. In the case of magnetite/hematite scale [17], the oxide scale is easy to crack in presence of Σ13b and Σ19c in α-Fe_2O_3 compared to Fe_3O_4 with Σ3. Thus, it is possible that during this time, tailoring specific grain boundaries can provide new insight into means of suppressing propagation of cracks when it is undesirable and into means of producing specific trapped nanoparticles when it is desired.

One thing we should consider is to distinguish grain boundary strengthening to steel substrate or to formed oxide scale itself. Extensive studies have been focused on the role of grain boundaries played in the steel substrate. For instance, grain boundary strengthening and precipitation hardening are considered to provide the most to the high-strength properties of the mechanically alloyed oxide dispersion strengthened (ODS) ferritic alloys, containing nano-sized (<3.5 nm) oxide dispersions. A higher density of these oxide particles with larger sizes than the ones in the matrix was found at the grain boundaries in ODS Fe–12Cr–5Al alloys (Y2O3 + ZrO2) (**Figure 6** [29]).

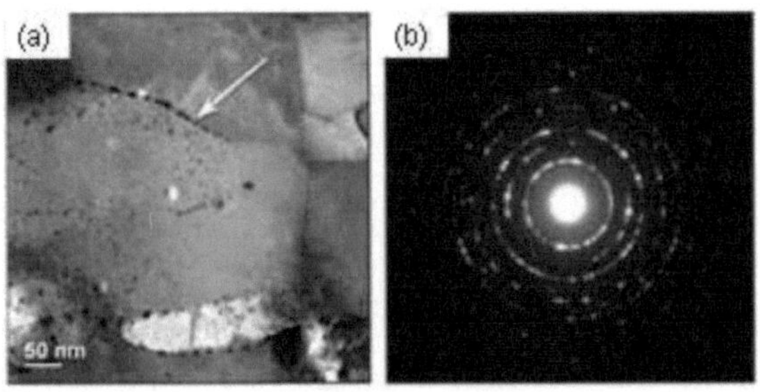

Figure 6. (a) High-angle annular dark field (HAADF)—scanning transmission electron microscopy (STEM) images of Fe–12Cr–5Al alloys (Y_2O_3 + ZrO_2) and (b) diffraction patterns after being mechanically alloyed and extruded at 950°C [29].

The grain refinement can be used to explain this strengthening process. Whatever grain boundaries works, the essential mechanism should be similar. However, the difference of a protective oxide scale containing reactive element may not simply be due to a site blocking effect in the grain boundary [30]. The contributions of dopant ions to the multiple electrical and ionic processes would provide valuable guidance to elucidate the deformation mechanism of oxide scale thermally formed on steel alloys. That is reason to note the difference between them as the presence of oxide particles at the grain boundaries and the temperature at which they were formed.

4.3. Effect of grain boundaries on tribological performance

Many different surface properties of metals and alloys will influence tribological performance. These surface properties include surface energy, crystallographic orientation, grain boundaries, texturing of surface and crystal structure. Grain boundaries in the oxide layers can alter the underlying failure mechanisms of formed oxide scale, which affects their tribological performance during metal processing. There are various strained conditions along grain boundaries because many dislocations present to help accommodate the misfit or mismatch

http://dx.doi.org/10.5772/66211

in adjacent orientations. These high energy regions at the surface could make sliding more difficult and increase the friction force of materials during metal forming.

Various mechanisms can be used to explain the role of grain boundaries in the tribolgoical properties of oxide scale during metal processing at high temperature. Our previous study [21] implies that grain boundary sliding contributes significantly to dissipation in oxide layers during hot rolling. If the oxidised grain boundary is under tension, both the metal and the oxide scale during thermal cycling tend to facilitate crack initiation [20]. A mechanism has been addressed for stress-aided grain boundary oxidation ahead of cracks. Oxygen embrittlement can therefore serve as the form of dynamic embrittlement or oxidation-induced grain boundary cracking during services at elevated temperatures [31]. In essence, the role of anisotropy needs to be investigated to clarify, which anisotropy (grain boundary energy or mobility) is dominant at which conditions. The local grain boundary planes can be dominated by the growing side of the boundary.

The friction characteristics with oxide scale also reveal a grain boundary effect—a profound dependence of friction on crystallographic direction and orientation and grain boundary characters. The variation of the coefficient of friction and rolling force in different thickness reductions during hot rolling associated with microtexture and grain boundary characters in magnetite/hematite scale formed on a microalloyed low carbon steel [17].

In summary, the role of grain boundary chemistry and structure on fundamental mechanisms and properties of oxide scale can help to accelerate the design optimisation of grain boundaries in oxidation or corrosion resistance.

5. Methodological potentials

Before starting, we need to consider some sample preparations to detect grain boundaries. Two directions are generally used to observe the oxide scale formed on the metal surface. One is the cross-sectional or thickness direction of the oxidised sample, parallel to the direction of oxide growth. Another is the top surface of oxidised sample suitable for the relatively thin oxide layers or the initial oxidation conditions. In order to visualise the grain boundary, the testing sample can be polished using mechanical and chemical-mechanical polishing methods. Particularly, it is necessary to select different etchants for different compositions in the oxide scale or metal substrate. Sometimes, this classical polishing/etching method cannot observe the grain morphologies both the oxide scale and the substrate concurrently. In electron backscattered diffraction (EBSD) technique and transmission electron microscope (TEM), the grain characters can be detected clearly without etching the sample. EBSD can use normal ion milling to prepare the sample. This is can make easier than TEM because TEM need to reduce the thickness of the sample using focus ion milling beam (FIB) or other advanced approaches. In a word, EBSD or TEM is generally used to observe the sample in cross-sectional direction, whereas scanning electron microscopy (SEM) can be used for top surface morphologies of oxidised sample. *In situ* electron microscopy cases are all the same but more challenging for the oxidation investigation.

Experimentally resolving and characterising grain boundary structure often requires a host of techniques: X-ray diffraction (XRD), scanning electron microscopy (SEM), electron backscattered diffraction (EBSD), transmission electron microscopy (TEM), X-ray energy dispersive spectroscopy (EDS) and electron energy loss spectroscopy (EELS). XRD or neutron diffraction normally deals with a texture that reflects an average value obtained from many different grains, that is, macrotexture. This chapter will address some techniques to obtain microtexture involving some individual grains.

5.1. Secondary electron microscope

Most grain boundary characters can be observed in low-vacuum secondary electron microscope (LV-SEM). In the backscattered electrons (BSE) mode, Z-contrast can assist phase identification. SEM/BSE hardly observe oxide scale and steel substrate without etching, because two parts of oxides and steel hardly to etching both using the same etchant. That is because polishing and etching for sample preparation can bring out the grain boundaries to more easily delineate individual grains. SEM/EDS and scanning transmission electron microscopy (STEM)/EELS help analyse the chemical species present and, combined with elemental mapping, can provide a distribution of the different chemistries in a spot, line or area. Sometimes, SEM can couple focus ion beam (FIB) to observe the grain characters along the cutting surface when preparing for TEM samples.

5.2. Electron backscattered diffraction EBSD

Electron backscattered diffraction (EBSD) can perform microstructure, phase identification, the crystallographic texture, and internal stresses, of oxidised sample. Some system can provide a transmission kikuchi diffraction (TKD) mode where the short working distance in backscattered electrons (BSE) detector as a complement. Various professional software suite fully integrated with image collection, versatile EBSD analysis and phase identification, can be used to acquire the online texture information and to analyse the offline scanning maps. Grain boundary mapping shows the crystallographic orientation of individual grains and the microstructure in the oxide scale [9]. Further, grain reconstruction can be carried out to delve the various mechanism associated with individual grains.

5.3. Transmission electron microscope

High-resolution transmission electron microscopy (HR-TEM) has improved microscopy resolution, more developed techniques, and coupling with advanced approaches will enable the understanding and engineering of grain boundaries (including twins) and intergranular films. For instance, scanning transmission electron microscopy (STEM) coupled with electron energy loss spectroscopy (EELS) is capable of simultaneously mapping the atomic/electronic structure of light elements such as oxygen at adequate spatial resolution. The electronic state of the elements across the boundary can be identified by STEM-EELS line scan crossing the grain boundary in steps of a few nanometres [32]. TEM imaging further resolves details of the crystal structure of grains and grain boundaries.

5.4. More advanced techniques

Atom probe tomography has some unique virtues for hydrogen detection, such as near-atomic resolution and equal sensitivity to all elements in the periodic table. The advanced technique has been used to investigate hydrogen embrittlement in the oxide scale of two common zirconium alloys [33]. Grain boundaries of oxides can be low-field areas of the APT specimen and then can readily be identified [34]. A combined use of TEM and APT can be used to quantify grain boundary segregation and has been applied to the case of carbon GB segregation in ferrite [24] and intergranular oxidation of a Ni–4Al alloy [35].

For some corrosion environment, electrochemical scanning tunnelling microscopy (ECSTM) has applied to analyse *in situ* the passivation of grain boundaries on microcrystalline copper in 0.1 M NaOH aqueous solution [36]. ECSTM has provided accurate *in situ* topographic information on grain and grain boundary effects on the local active dissolution of microcrystalline copper in acid solution.

Time-of-flight secondary ion mass spectrometry (TOF-SIMS) provides elemental, chemical state and molecular information from surfaces of solid materials. Analogous to SEM/EDS instruments, TOF-SIMS aims to the compositional analysis of ultra-thin layers and nanoscale sample features. The difference is that TOF-SIMS can be used to characterise molecular information from organic materials and tissue sections for medical research.

6. Conclusions and future directions

This chapter covers the recent advance associated with grain boundaries in the oxide scale formed on metal alloys during metal processing. A number of benefits include (i) characterisation of grain boundaries ranging from microstructure, preferred orientations and different types of grain boundaries; (ii) the role of grain boundaries in the oxide scale playing in diffusion-controlled oxidation, deformation mechanism and tribological performance; and (iii) introducing the experimental techniques and analytical methodology underpinning this subject.

Some specific results can be concluded ranging from micotexture and grain boundaries characters. The (100) plane of magnetite is much more sensitive to the oxidation. In the coincident site lattice (CSL) boundaries, the $\sum 3$ in magnetite and $\sum 13b$ in hematite are dominant in the oxide scale. These findings suggest that low-angle grain boundaries and low-energy CSL boundaries can be used to prevent the initiation and propagation of cracks, further to enhance oxidation resistance of the materials.

Three current challenges dominate in the characterisation, mechanism and techniques for investigation the grain characters in oxide scale during metal processing at high temperature. First, grain characters consisting of grain shape and size, phase grain boundaries within oxide scale and orientation relationship between oxides are also need to be considered. Second, to delve which types of (special) grain boundaries to enhance the oxidation/corrosion resistance and then to tailoring them. Finally, a combination of the various advanced techniques provides the frameworks for future investigation on the oxidation of the other metal alloys even bulk ceramics.

Acknowledgements

This work was supported by the China Postdoctoral Science Foundation under Grant No. 2015M580094, and the National Natural Science Foundation of China under Grant Nos.11274198 and 51532004.

Author details

Xianglong Yu* and Ji Zhou

*Address all correspondence to: xly991@uow.edu.au

State Key Laboratory of New Ceramics and Fine Processing, School of Materials Science and Engineering, Tsinghua University, China

References

[1] Birks N, Meier GH. Introduction of High Temperature Oxidation of Metals. 2nd ed. London: Cambridge University Press; 2006.

[2] Comel RM, Schwertmann U. The Iron Oxides: Structure, Properties, Reactions, Occurrence and Uses. New York: Wiley; 2003.

[3] Yu X, Jiang Z, Zhao J, Wei D, Zhou C, Huang Q. Microstructure and microtexture evolutions of deformed oxide layers on a hot-rolled microalloyed steel. Corrosion Science. 2015;**90**:140–152. doi:10.1016/j.corsci.2014.10.005

[4] Chen RY, Yuen WYD. Review of the high-temperature oxidation of iron and carbon steels in air or oxygen. Oxidation of Metals. 2003;**59**:433–468. doi:10.1023/A:1023685905159

[5] Saunders SRJ, Monteiro M, Rizzo F. The oxidation behaviour of metals and alloys at high temperatures in atmospheres containing water vapour: a review. Progress in Materials Science. 2008;**53**:775–837. doi:10.1016/j.pmatsci.2007.11.001

[6] Juricic C, Pinto H, Cardinali D, Klaus M, Genzel C, Pyzalla AR. Effect of substrate grain size on the growth, texture and internal stresses of iron oxide scales forming at 450°C. Oxidation of Metals. 2010;**73**:15–41. doi:10.1007/s11085-009-9162-1

[7] Yu X, Jiang Z, Zhao J, Wei D, Zhou C, Huang Q. Effect of a grain-refined microalloyed steel substrate on the formation mechanism of a tight oxide scale. Corrosion Science. 2014;**85**:115–125. doi:10.1016/j.corsci.2014.04.006

[8] Samal S, Mitra SK. Influence of grain shape, size, and grain boundary diffusion on high-temperature oxidation of pure metal Fe, Cu, and Zn. Metallurgical and Materials Transactions A. 2015;**46**:3324–3332. doi:10.1007/s11661-015-2987-0

[9] Engler O, Randle V. Introduction to Texture Analysis: Macrotexture, Microtexture and Orientation Mapping. 2nd ed. Boca Raton: CRC Press; 2010.

[10] Higginson RL, Roebuck B, Palmiere EJ. Texture development in oxide scales on steel substrates. Scripta Materialia. 2002;**47**:337–342. doi:10.1016/S1359-6462(02)00154-9

[11] Suárez L, Rodríguez-Calvillo P, Houbaert Y, Garza-Montes-de-Oca NF, Colás R. Analysis of deformed oxide layers grown on steel. Oxidation of Metals. 2011;**75**:281–295. doi:10.1007/s11085-010-9231-5

[12] Yu X, Jiang Z, Zhao J, Wei D, Zhou C, Huang Q. Crystallographic texture based analysis of Fe_3O_4/α-Fe_2O_3 scale formed on a hot-rolled microalloyed steel. ISIJ International. 2015;**55**:278–284. doi:10.2355/isijinternational.55.278

[13] Chang L, Lin SN. Analytical electron microscopy study of interfacial oxides formed on a hot-rolled low-carbon steel. Oxidation of Metals. 2005;**63**:131–144. doi:10.1007/s11085-004-3196-1

[14] Deepak K, Mandal S, Athreya CN, Kim D, de Boer B, Sarma S. Implication of grain boundary engineering on high temperature hot corrosion of alloy 617. Corrosion Science. 2016;**106**:293–297. doi:10.1016/j.corsci.2016.01.019

[15] Palumbo G, Aust KT, Lehockey EM, Erb U, Lin P. On a more restrictive geometric criterion for special CSL grain boundaries. Scripta Materialia. 1998;**38**:1685–1690. doi:10.1016/S1359-6462(98)00077-3

[16] Lehockey EM, Brennenstuhl AM, Thompson I. On the relationship between grain boundary connectivity, coincident site lattice boundaries, and intergranular stress corrosion cracking. Corrosion Science. 2004;**46**:2383–2404. doi:10.1016/j.corsci.2004.01.019

[17] Yu X, Jiang Z, Zhao J, Wei D, Zhou C, Huang Q. Effects of grain boundaries in oxide scale on tribological properties of nanoparticles lubrication. Wear. 2015;**332–333**:1286–1292. doi:10.1016/j.wear.2015.01.034

[18] Phaniraj MP, Kim DI, Cho YW. Effect of grain boundary characteristics on the oxidation behavior of ferritic stainless steel. Corrosion Science. 2011;**53**:4124–4130. doi:10.1016/j.corsci.2011.08.020

[19] Yamaura S, Igarashi Y, Tsurekawa S, Watanabe T. Structure-dependent intergranular oxidation in Ni–Fe polycrystalline alloy. Acta Materialia. 1999;**47**:1163–1174. doi:10.1016/S1359-6454(99)00007-5

[20] Fedorova E, Braccini M, Parry V, Pascal C, Mantel M, Roussel-Dherbey F, Oquab D, Wouters Y, Monceau D. Comparison of damaging behavior of oxide scales grown on austenitic stainless steels using tensile test and cyclic thermogravimetry. Corrosion Science. 2016;**103**:145–156. doi:10.1016/j.corsci.2015.11.012

[21] Yu X, Jiang Z, Zhao J, Wei D, Zhou J, Zhou C, Huang Q. Dependence of texture development on the grain size of tertiary oxide scales formed on a microalloyed steel. Surface and Coatings Technology. 2015;**272**:39–49. doi:10.1016/j.surfcoat.2015.04.026

[22] Horita T, Kishimoto H, Yamaji K, Xiong Y, Sakai N, Brito ME, Yokokawa H. Effect of grain boundaries on the formation of oxide scale in Fe–Cr alloy for SOFCs. Solid State Ionics. 2008;**179**:1320–1324. doi:10.1016/j.ssi.2008.01.090

[23] Yan J, Gao Y, Gu Y, Sun F, Yang Z, Lu J, Yin H, Li Y. Role of grain boundaries on the cyclic steam oxidation behaviour of 18-8 austenitic stainless steel. Oxidation of Metals. 2016;**85**:409–424. doi:10.1007/s11085-015-9603-y

[24] Kruska K, Lozano-Perez S, Saxey DW, Terachi T, Yamada T, Smith GDW. Nanoscale characterisation of grain boundary oxidation in cold-worked stainless steels. Corrosion Science. 2012;**63**:225–233. doi:10.1016/j.corsci.2012.06.030

[25] Kim JH, Kim DI, Shim JH, Yi KW. Investigation into the high temperature oxidation of Cu-bearing austenitic stainless steel using simultaneous electron backscatter diffraction-energy dispersive spectroscopy analysis. Corrosion Science. 2013;**77**:397–402. doi:10.1016/j.corsci.2013.08.015

[26] Dugdale H, Armstrong DEJ, Tarleton E, Roberts SG, Lozano-Perez S. How oxidized grain boundaries fail? Acta Materialia. 2013;**61**:4707–4713. doi:10.1016/j.actamat.2013.05.012

[27] Heuer AH, Azar MZ. A disconnection mechanism of enhanced grain boundary diffusion in Al_2O_3. Scripta Materialia. 2015;**102**:15–18. doi:10.1016/j.scriptamat.2015.01.026

[28] Jiang R, Gao N, Reed PAS. Influence of orientation-dependent grain boundary oxidation on fatigue cracking behaviour in an advanced Ni-based superalloy. Journal of Materials Science. 2015;**50**:4379–4386. doi:10.1007/s10853-015-8992-2

[29] Unocic KA, Pint BA, Hoelzer DT. Advanced TEM characterization of oxide nanoparticles in ODS Fe–12Cr–5Al alloys. Journal of Materials Science. 2016;**51**:9190–9206. doi:10.1007/s10853-016-0111-5

[30] Rettberg LH, Laux B, He MY, Hovis D, Heuer AH, Pollock TM. Growth stresses in thermally grown oxides on nickel-based single-crystal alloys. Metallurgical and Materials Transactions A. 2016;**47**:1132–1142. doi:10.1007/s11661-015-3273-x

[31] Chan KS. A grain boundary fracture model for predicting dynamic embrittlement and oxidation-induced cracking in superalloys. Metallurgical and Materials Transactions A. 2015;**46**:2491–2505. doi:10.1007/s11661-015-2860-1

[32] Lin Y, Fang S, Su D, Brinkman KS, Chen F. Enhancing grain boundary ionic conductivity in mixed ionic-electronic conductors. Nature Communications. 2015;**6**:6824. doi:10.1038/ncomms7824

[33] Sundell G, Thuvander M, Yatim AK, Nordin H, Andrén HO. Direct observation of hydrogen and deuterium in oxide grain boundaries in corroded Zirconium alloys. Corrosion Science. 2015;**90**:1–4. doi:10.1016/j.corsci.2014.10.016

[34] Kim JH, Kim BK, Kim DI, Choi PP, Raabe D, Yi KW. The role of grain boundaries in the initial oxidation behavior of austenitic stainless steel containing alloyed Cu at 700°C for advanced thermal power plant applications. Corrosion Science. 2015;**96**:52–66. doi:10.1016/j.corsci.2015.03.014

[35] Schreiber DK, Olszta MJ, Bruemmer SM. Directly correlated transmission electron microscopy and atom probe tomography of grain boundary oxidation in a Ni–Al binary alloy exposed to high-temperature water. Scripta Materialia. 2013;**69**:509–512. doi:10.1016/j.scriptamat.2013.06.008

[36] Chen H, Maurice V, Klein LH, Lapeire L, Verbeken K, Terryn H, Marcus P. Grain boundary passivation studied by *in situ* scanning tunneling microscopy on microcrystalline copper. Journal of Solid State Electrochemistry. 2015;**19**:3501–3509. doi:10.1007/s10008-015-2787-x

Chapter 5

Grain Boundary Segregation in Nanocrystructured Metallic Materials

Jinyu Zhang, Gang Liu and Jun Sun

Additional information is available at the end of the chapter

http://dx.doi.org/10.5772/66598

Abstract

The aim of this chapter is to shed light on the effects of grain boundary segregation on microstructural evolution in nanostructured metallic materials as well as on their mechanical properties. Several key topics will be covered. First, a brief explanation of mechanical stress-driven grain growth in nanostructured Al, Ni, and Cu thin films will be provided in terms of a deformation mechanism map. It will become clear that the excess energy of grain boundaries enable the nanostructured metals to suffer from significant microstructure evolution via dislocation-boundary interactions during plastic deformation even at room temperature. Manipulation of grain boundary structures/properties via dopants segregation at grain boundaries to inhibit grain coalescence associated with remarkably enhanced mechanical properties is then discussed in three representative binary Cu-based systems, i.e., Cu-Zr, Cu-Al, and Cu-W. This is finally followed by a summary of this chapter.

Keywords: nanostructured materials, mechanical properties, grain boundary segregation, grain growth, twinning

1. Introduction

The nanoscaled internal features, in which two quantities, i.e., the *characteristic length* and the *size parameter* likely overlap, render the conventional deformation mechanisms and size laws often break down and even be reversed in nanostructured (NS) metallic materials, including nanocrystalline (NC) and nanotwinned (NT) materials [1–3]. Indeed, for many purposes, making the crystals as small as possible provides significant advantages in performance, but such materials are often unstable: The crystals tend to merge and grow larger even at room temperature (RT) [4–7]. To suppress the nanograin growth and maintain the nanoscale microstructure, two approaches can be adopted [8–15]: (i) Thermodynamic approach: reduction of the driving force for grain growth by segregation of the solutes at the grain-boundaries (GBs), (ii) Kinetic approach: reduction of the GB mobility, by e.g., porosity, solute atoms, and precipitates at the GBs, which impose drag forces.

In this chapter, first, a brief introduction of mechanical stress-driven grain growth in NS Cu and Ni thin films/foils as well as their mechanical properties will be provided in terms of size-dependent deformation mechanisms. Subsequently, dopants segregation at GBs to hinder grain coarsening and enhance mechanical properties via the alloying method is discussed in three representative binary Cu-based systems, i.e., Cu-Zr, Cu-Al, and Cu-W.

2. Synthesis, microstructural characterizationw and mechanical tests

The synthesis of NS metallic thin films can be achieved by several bottom-up techniques, such as physical vapor deposition (PVD) and electrodeposition (ED), in which the choice of deposition conditions has a tremendous influence on the microstructural features and mechanical properties of these NS metallic films/foils.

PVD is the most common approach to fabricate metallic thin films/foils, including evaporation, sputtering, and less commonly molecular beam epitaxy [16, 17]. Compared with other methods, magnetron sputtering (MS) can clean the substrate by "backsputtering" and generate greater impact angles of the sputtered atoms onto the substrate, resulting in smaller surface roughness of the film by covering the defects and/or step on the substrate [17]. Although, MS increases the possibility of crystal damage due to high impact energies of sputtered atoms, it is still the most widely used method to prepare thin films.

ED is a technique within the broader group of electrochemical synthesis methods and uses an electric current to deposit pure metals from an aqueous, electrolytic solution [18, 19]. Compared with PVD, ED offers a lower cost and faster low-temperature deposition method. It displays remarkable advantages to synthesize highly dense NC materials with (1) few size and shape limitations, (2) tunable microstructural size parameters, and (3) hierarchical structures, e.g., a bimodal grain size-distribution [20] and NT grains [21], providing potential benefits to mechanical performance. Especially, these nanotwins improve both the mechanical strength and ductility, yet maintain high electric conductivity [22].

The crystalline structure, orientation, and grain boundaries within metallic thin films could be experimentally probed by suitable techniques, including X-ray diffraction (XRD), scanning and transmission electron microscopy (SEM and TEM), combining with other more superior appurtenances, such as the electron backscattered diffraction (EBSD) system and the precession-enhanced electron diffraction (PED) system. The chemical conuration of the materials can be characterized by the energy dispersive X-ray (EDX) and the powerful 3-D atom probe tomography (APT).

Due to the difficulty in performing the mechanical tests on the free-standing metallic thin films often with thickness of roughly 1 μm or less, researchers put great emphasis on the substrate-supported thin films. For example, the tensile ductility and fatigue lifetime of metallic thin films on flexible substrates, both of which are characterized by the critical strain to nucleate microcracks [23, 24], can be determined by a Micro-Force Test System (MTS® Tytron 250) at RT. By contrast, the strength/hardness and modulus of thin films on rigid substrates can be measured using instrumented nanoindenter apparatus (e.g., TI950 TriboIndenter, Nano XP) often equipped with a standard Berkovich tip and a diamond flat punch. In what follows, we will mainly concentrate on the mechanical properties of substrate-supported metallic (alloyed) NS thin films.

3. Microstructure-mechanical properties correlation of nanostructured pure FCC metals

This section is divided into three subsections. The subsection on size-dependent deformation mechanisms is introduced based on a deformation-mechanism map. Microstructural evolution, in particular, the steady-state grain size, is then discussed in terms of a dislocation-based mechanism. The mechanical properties subsection contrasts yield strength, ductility, strain-rate sensitivity, and fatigue lifetime in NS metals.

3.1. Size-dependent deformation mechanisms

In coarse-grained (CG) metals (grain size $d \geq 1$ μm), deformation of the material is believed to occur through the generation and motion of dislocations within the individual grains. As the grain size decreases, it is normally expected that with GBs now occupying a significant volume fraction of the material, deformation proceeds by a mechanism that is intergranular rather than intragranular in nature.

Yamakov and colleagues [25] constructed a deformation mechanism map in NC FCC metals using information obtained from molecular dynamics (MD) simulations (see **Figure 1**), revealing how the crossover with decreasing d from dislocation-driven to GB-mediated deformation depends on the stacking-fault energy (SFE, γ_{sf}), the elastic properties of the material, and the magnitude of the applied stress. This deformation map can be divided into three regions in light of the competition between the grain size d and the dislocation splitting distance r. Region I encompasses larger d and/or higher γ_{sf}, where plastic deformation is dominated by full (perhaps extended) dislocations that nucleate from GBs and propagate across grains. Region II involves smaller d and/or lower γ_{sf}, where partials nucleate and propagate across grains, associated with production of stacking-faults (SFs) that inhibit subsequent dislocation motion and induce strain hardening. Region III corresponds to the smallest d or lowest stress regime, where no dislocations are present and deformation is dominantly controlled by GB-mediated mechanism, resulting in an inverse H-P effect. Although these MD simulations performed at unrealistically large strain rates (10^7–10^9 s^{-1}), their findings agree well with the experiment observations about the transition from full dislocations to partials and finally to GB-mediated processes.

Where do dislocations in NC metals go if they are the dominant plastic carriers? Actually, in the MD simulations, a key deformation process is dislocation nucleation at a GB, glide across grain interiors that are free of obstacles, and are absorbed by the opposite GB [26, 27].

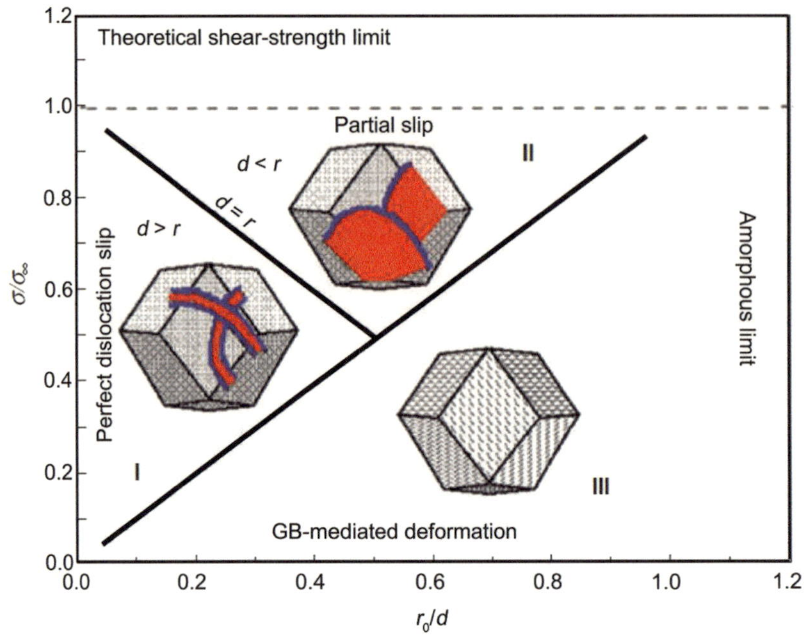

Figure 1. A deformation-mechanism map incorporating the role of the SEF for FCC NC metals at low temperature. The map shows three distinct regions in which either complete extended dislocations (Region I) or partial dislocations (Region II), or no dislocations at all (Region III) exist during the low-temperature deformation of FCC NC metals. The map is expressed in reduced units of stress (σ/σ_∞) and inverse grain size (r_0/d).The parameters σ_∞ and r_0 are functions of the SEF and the elastic properties of the material. Figure is taken with permission from Ref. [25].

Both *in situ* and *ex situ* experiments have unambiguously demonstrated that the reversible motion of dislocations emitted from GBs accompanied with weak storage even starvation of dislocations in grain interiors in deformed NC FCC metals, leaving behind deformation twins (terminated at GBs) and SFs in general at upper and lower nanoregime, respectively [28–34]. In this context, GBs act as dislocation sources as well as sinks.

In parallel, several theoretical models have been proposed to predict the crossover grain size (d_C) between emitted full dislocations and partials from GBs and explain the twinning behavior in NC metals. A simple, realistic model based on dislocations emission from GBs was constructed by Asaro et al. [35], in which the critical stresses needed to move a full dislocation and a partial are described respectively as follows:

$$\sigma_{\text{Full}} \propto \frac{\mu b}{d} \tag{1}$$

and

$$\sigma_{\text{Partial}} \propto \left(\frac{1}{3} - \frac{1}{12\pi}\right)\frac{\mu b}{d} + \frac{\gamma_{SF}}{b} \tag{2}$$

where μ is the shear modulus, b is the magnitude of Burgers vector of full dislocations, and γ_{SF} is the stacking fault energy (SFE).

The slip of partials in general triggers the formation of deformation twins and SFs that contribute to the plastic deformation of NC FCC metals. There is a double-inverse grain size effect on deformation twinning in NC metal with respect to the normal Hall-Petch (H-P) *d*-dependence, as uncovered in Ni [30] and Cu [31]. This nonmonotonic *d*-dependence of twinning is explained by Zhang et al. [31] via the stimulated slip model, involving the competing grain size effects on the emission of the first partial, and the plane-to-plane promotion of partial slip afterwards. Though this model was originally proposed to explain the H-P *d*-dependent twinning in CG metals (e.g., Ti), latter *in situ* TEM observations in stretched Au nanowires clearly demonstrated the stimulated slip of partials is operative at nanoregime [36]. Actually, just opposite to the trend of twinning, its reverse process, i.e., detwinning, also manifests the double-inverse *d*-dependence, as revealed in NT Ni [37].

It is conceivable that GB-mediated deformation become more important in NC metals due to a high density of GBs [38, 39]. Typically, this is expected to occur for grain sizes below 15 nm for most metals [38], because ordinary dislocation plasticity requires prohibitively high stresses to switch on, predicted from Eqs. (1) and (2). In this regime, GB-mediated deformation leads to material's softening or the so-called inverse H-P effect [40]. Given the pervasive dislocation nucleation and motion still prevails in such a small size-range, Carlton and Ferreira [41] established an elegant model based on the statistical absorption of dislocations by GBs to explain the inverse H-P effect, showing that the yield strength is dependent on strain rate and temperature and deviates from the H-P relationship below a critical grain size.

Building on these insights from NC metals, it is unexpected that Cu, even high SFE Ni, with submicron grains and a high density of nm-scale twin boundaries (TBs) exhibit the softening behavior deformed at RT. As a matter of fact, in NT FCC metals, the TBs not only serve as deformation barrier for dislocation transmission but also serve as dislocation sources as well as sinks [21, 42, 43]. Concomitantly, NT metals, e.g., Cu, also exhibit the size-dependent deformation mechanisms that transit from dislocation nucleation from steps on the TBs to TB/GB junctions at a critical twin thickness (λ_C), e.g., λ_C~18 nm for Cu [43], as shown in **Figure 2**. At this point, the classical H-P type of strengthening due to dislocation pile-up and cutting through twin planes transforms to a dislocation-nucleation controlled softening mechanism with TB migration resulting from nucleation and motion of partials parallel to the twin planes [44]. This mechanism transition size is quantitatively consistent well with the strongest size of ~15 nm determined by mechanical tests [21].

To summarize, NC metals exhibit size-dependent deformation mechanisms at different size regimes that involve GBs as the primary sources and sinks for dislocations as well as diffusive and sliding phenomena, that is to say, the size-dependence itself manifest strong size effects. This would inevitably affect the microstructural evolution and mechanical properties addressed below.

3.2. Microstructure evolution in nanostructured metals

Understanding the underlying physical mechanisms of grain growth/refinement in materials, in particular, for NT metals with simultaneous high strength and good ductility, to manipulate their microstructural stability for performance optimization is a grand challenge in the material community. It is well realized that the CG metals would shrink their grains, whereas

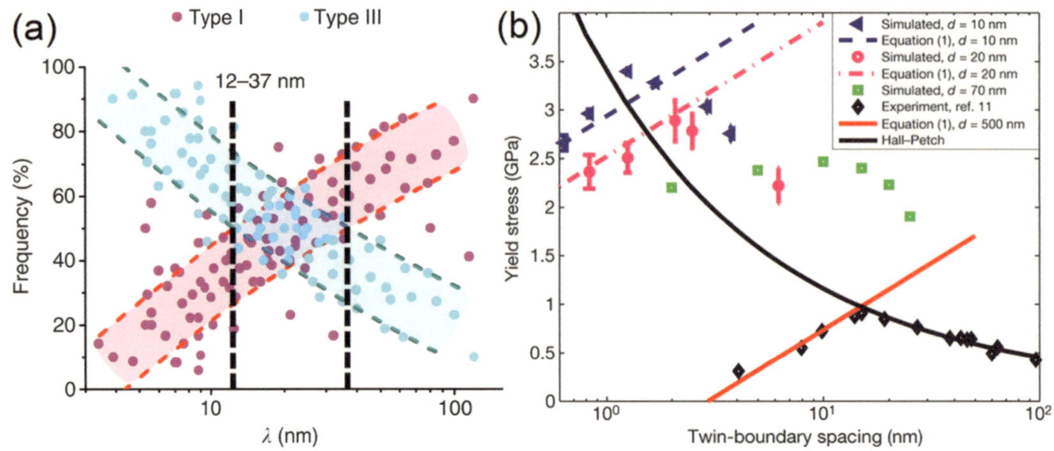

Figure 2. (a) Statistical distribution of two types of dislocations in NT Cu with different TB spacing λ during *in situ* deformations (left, figure is taken with permission from Ref. [43]). (b) Yield stress of NT Cu as a function of TB spacing λ at different grain sizes (right, figure is taken with permission from Ref. [44]).

the NC metals often coarsen their grains during plastic deformation even at low temperature. Similar phenomena were observed in NT metals and alloys, such as Cu. Therefore, it is naturally anticipated that for a metal it has a steady-state grain size (d_S) during plastic deformation, which was previously taken granted as a characteristic of each metal.

3.2.1. Steady-state grain size

So far, the steady-state grain sizes of metals have been thoroughly modeled in terms of various physical parameters by Mohamed [45] and further analyzed by Edalati and Horita [46] with respect to atomic bond energy and related parameters. The usage of applied stress (σ_a) in Mohamed's model [45] renders the roles of average internal stresses (σ_i) driving recovery or average effective stresses ($\sigma_e = \sigma_a - \sigma_i$) driving dislocation motion played in microstructural evolution during plastic deformation are indistinguishable. This treatment would miss some critical information about the physical mechanism(s) for microstructural evolution, which is unfavorable for us to design an engineering material with the steady-state grain size via tuning their initial microstructures and/or processing parameters. In the light of competition between average effective and internal stresses characterized by the stress ratio $\eta_{\text{Stress}} = \sigma_e/\sigma_i$, Li and coworkers [47] most recently constructed a new dislocation-based model to describe the steady-state grain size d_S for NS metals as

$$\frac{d_s}{b} = C\left[\frac{(2+\upsilon)M}{162\pi K b^2 \omega^2 \varphi}\right]\left(\frac{\mu b}{\gamma_{sf}}\right)\left(\frac{\mu}{\sigma_e}\right) \quad \text{for } (f = 0) \tag{3}$$

and

$$\frac{d_s}{b} = \frac{\sqrt{(1-f)^2 + C\frac{f}{\lambda}\frac{(4+2\upsilon)M}{81\pi K b \omega^2 \varphi}\left(\frac{\mu b}{\gamma_{sf}}\right)\left(\frac{\mu}{\sigma_e}\right)}}{2f}\frac{\lambda}{b} - \frac{1-f}{2f}\frac{\lambda}{b} \quad \text{for } (0 < f \le 1) \tag{4}$$

where C is a stress-dependent coefficient in-between the growth rate C_1 and the refinement rate C_2, and a useful representation of the coefficient C as a function of σ_e, consisting of C_1 and C_2 below and above the internal stress σ_i, respectively, is $C = ((C_1+C_2)/2)-((C_1-C_2)/2)\mathrm{erf}((\sigma_e-\sigma_i)/\Delta)$, Δ is a measure of the extent of the transition region, M is the Taylor factor, μ is the shear modulus, υ is Poisson's ratio, φ *is* the misorientation angle between neighboring grains, ω represents the linear atomic density of the dislocation line, f is the number fraction of nanotwins, and K is a constant ($K = 1$ for screw dislocations and $K = (1 - \upsilon)$ for edge dislocations). This model captures well with the steady-state grain size d_S obtained from free-standing NS Ni foils with $\lambda = 38$ nm at the steady-state creep stage tested at RT, as shown in **Figure 3**. Interestingly, by postmortem transmission electron microscopy (TEM) observations, Li et al. [47] uncovered that the ED NT Ni foils (with a strong (111) peak and followed by (200) and (311) peaks) prefer to display grain coalescence at low stress ratios $\eta_{\mathrm{Stress}} < 1$, while they prefer to display grain refinement at stress ratios $\eta_{\mathrm{Stress}} > 1$ during the creep test. When the effective stress balances the internal stress, i.e., $\eta_{\mathrm{Stress}} = 1$, these NT Ni foils sustain the stable microstructures. Note that the stress ratio itself is strongly temperature- and strain rate-dependent, in that the internal stress σ_i has contained the contribution of the thermal component. In their work, the grain refinement/growth in the NT Ni is achieved by twinning- or detwinning-mediated mechanism via dislocation-boundary interactions. Similar phenomena in cyclically compressed bulk NC Cu with an initial d of ~25 nm and in fatigued ultrathin Au thin films with an initial d of ~19 nm were observed by Hu et al. [34] and Luo et al. [48], respectively, at RT. The underlying reasons for grain growth are the excessive energy of GBs/TBs and randomly orientated grains [49] in these Ni and Cu nanostructures synthesized by the nonequilibrium deposition.

3.2.2. *Mechanisms of grain growth and grain refinement*

Traditionally, mechanistic descriptions that have been developed to describe NC metals have generally considered the GBs to be stable and immortal obstacles to dislocation motion, whereas there are numerous evidences that suggest that this is not always the case [4–7]. Such materials are often unstable: The NC grains tend to merge and grow larger as subjected to heat or stress. Indeed, *in situ* nanoindentation of ultrafine-grained (UFG)/NC Al films deposited on specially designed Si wedges demonstrated rapid GB migration and coalescence during deformation [4]. Another representative study has reported the grain growth of UFG/NC Cu near the indented region during microhardness testing at both cryogenic temperature and RT by Zhang et al. [5]. They surprisingly uncovered that the grain growth was found to be faster at cryogenic temperature than at RT, implying that the grain coarsening process is driven primarily by stresses rather than diffusion. Gianola et al. [6, 7] concluded that stress-driven grain growth appears to have preceded dislocation activity and involved GB migration and grain coalescence and becomes an active RT deformation mode in abnormally ductile NC Al thin films, based on coupled microtensile thin-film testing, *in situ* synchrotron diffraction experiments, and postmortem TEM observations. However, the twinning-mediated grain grow mechanism unveiled in stretched NT Ni [37, 47] and compressed NT Cu [34] is radically different from these grain growth mechanisms aforementioned above.

Figure 4 shows the atomic evidence of twinning-mediated grain growth in NT Ni, essentially being the consequence of nanotwin-assisted GB dissociation and local grain coarsening. Because the localized misorientation between two adjacent grains G1 and G2 can be reduced

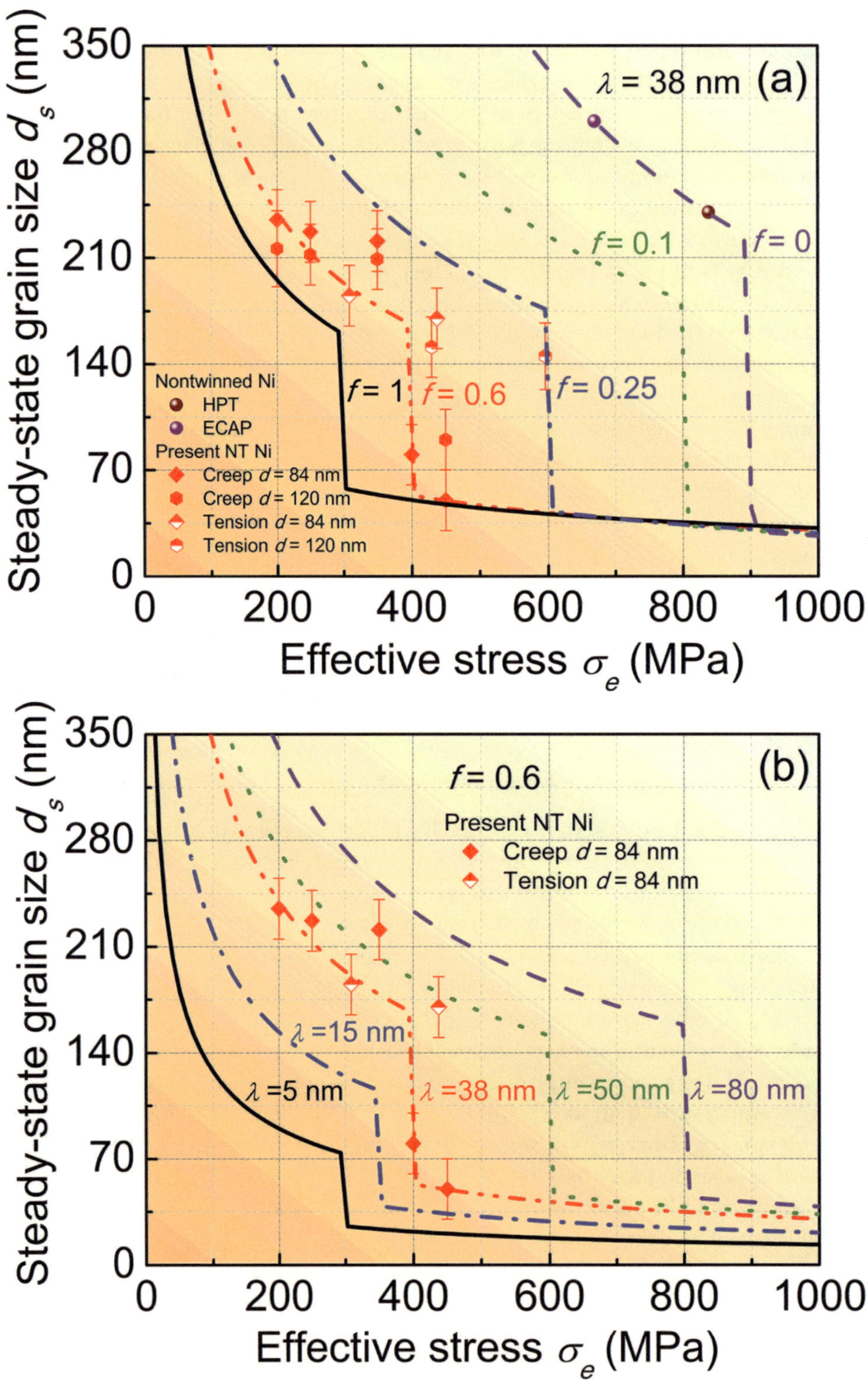

Figure 3. (a) Calculated steady-state grain size ds as a function of average effective stress σ_e in nanograined Ni with different twins' fraction f (a) and twin thickness λ (b). Figure is taken with permission from Ref. [47].

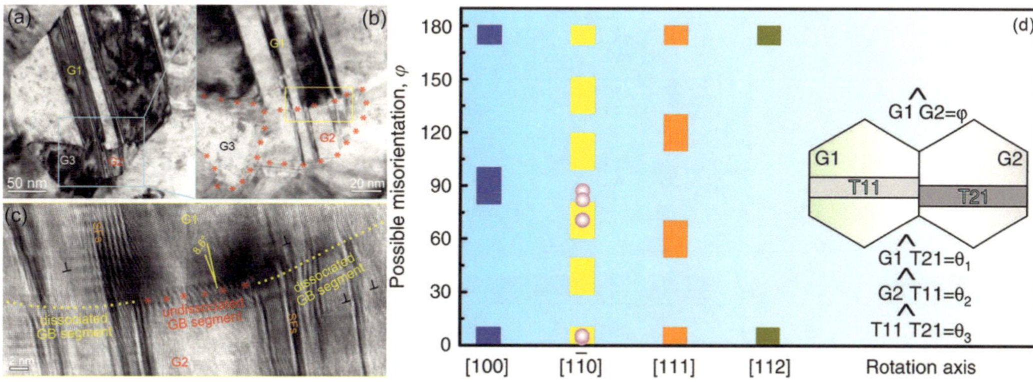

Figure 4. The TEM images showing nanotwin-assisted grain growth occurred among three grains (labeled as G1, G2 and G3, respectively) in NT Ni after creep. (b) is the magnified view of blue rectangular region in (a) and (c) is the magnified view of yellow rectangular region in (b) showing grain coalescence between G1 and G2 (left, figure is taken with permission from Ref. [47]). (d) Possibility for nanotwin-assisted grain coalescence. Possible misorientation angle (ϕ) suitable for nanograins G1 and G2 coalescence induced by the present nanotwin-assisted mechanism under different rotation axes <*hkl*> (right, figure is taken with permission from Ref. [48]). The inset shows all the misorientation angles among grains and twins.

by twinning, leading to some parts of G1 being transformed into G2, thus some localized segments of GB coalesce and disappear, see **Figure 4(b)** and **(c)**. Consequently, the repetitive formation of nanotwins induces some local segments of a high-angle GB are transformed into low-angle GB segments by storage of residual dislocations generated from dislocation reactions [37]. These recurrent interactions between partials/twins and GBs would facilitate the two adjacent nanograins to gradually coalesce into one larger grain with nanotwins. Moreover, there is a great possibility for the present mechanism to occur in G1/G2 with different mutual misorientation (φ) through the rotation around four typical low-index symmetric axes <*hkl*>, in particular, for φ<*111*> (φ = 0–10°, 50–70°, 110–130°, and 170–180°) and φ<*110*> (φ = 0–10°, 29–48.9°, 60.6–80.5°, 99.5–119.4°, 131.1–151°, and 170–180°) [48], as shown in **Figure 4(d)**.

Figure 5 displays the TEM observation of detwinning-induced refinement of grains in NT Ni, achieved by interplay between partials and primary TBs. Two typical examples of the interactions are presented in **Figure 5(b)** and **(c)**. It appears that the atomic arrangement is distorted at the intersection region of twins, see **Figure 5(b)**. The presence of SFs in the primary twin implies the gliding of partials created by dislocation-TB reactions [50]. In **Figure 5(c)**, these Shockley partials glide parallel to the CTB, rendering detwinning of the primary twin, as observed in the twins crossed region of R4. As deformation proceeds, these partials stimulate twinning process, resulting in twin interactions to produce abundance of sessile dislocations. As a consequence, CTBs lose their coherency and transform into conventional GBs [51]. Obviously, this mechanism is parallel with other mechanisms for nanoscale structural refinement via twin/matrix lamellae in various FCC metals are identified, such as fragmentation of T/M lamellae, twins intersection, and shear banding [52].

3.3. Size effects on the mechanical properties

Mechanical properties of nanoscale structures are well known for deviating from their CG counterparts, exhibiting size effects across a wide range of properties. These NS metallic

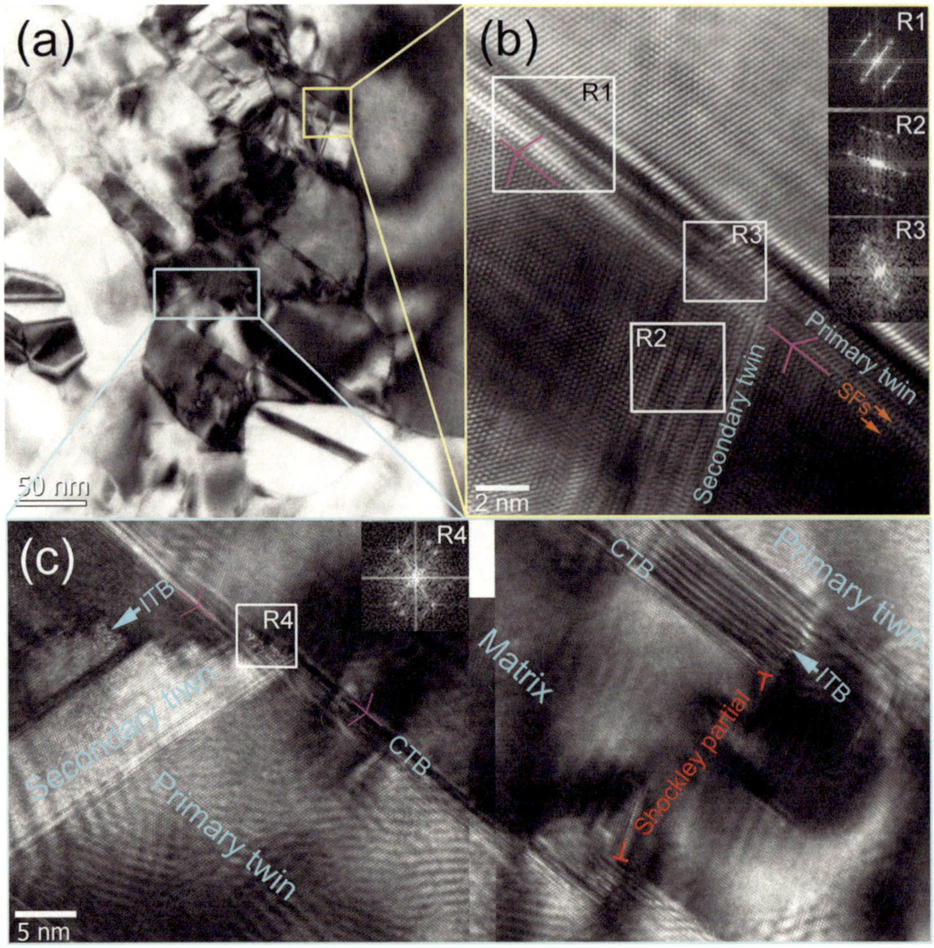

Figure 5. The TEM images showing the GB formation for grain refinement via detwinning-induced twin interactions in NT Ni after creep. Figure is taken with permission from Ref. [47].

materials generally fall under the banner of "smaller is stronger." The result of this size effect is that NS thin films often exhibit mechanical properties of an increased magnitude: typically the yield strength, strain rate sensitivity (SRS), and fatigue lifetime all increase with respect to the accepted bulk values.

3.3.1. Strength and ductility

A striking feature of NS metals is their extraordinary strength compared to corresponding bulk materials. The dependence of measured yield strength σ_y of either substrate supported [23, 53–57] or freestanding [56] Cu on film thickness h and on grain size d are summarized and shown in **Figure 6(a)** and **(b)**, respectively. It seems that, similar to their bulk NS counterparts, σ_y of Cu thin films also monotonically increases with reducing d and shows some-

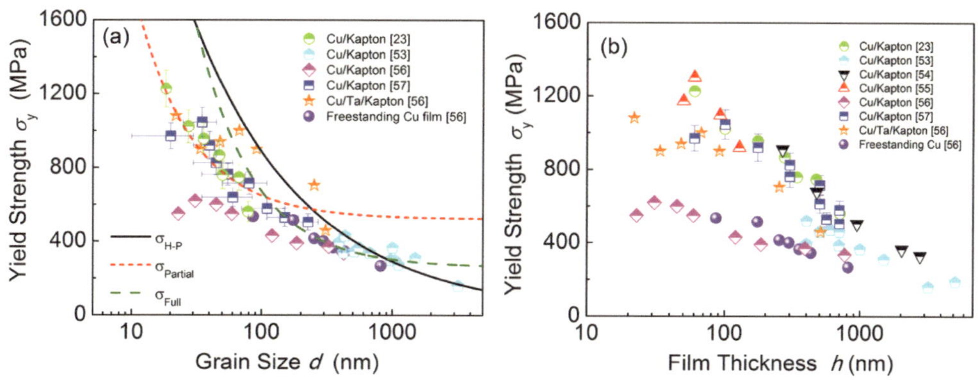

Figure 6. The dependence of yield strength of Cu thin films as a function of (a) grain size *d* and (b) film thickness *h*. The lines in (a) are predictions of yield strength from the H-P relationship, partial dislocation model (Eq. (1)) and full/perfect dislocation model (Eq. (2)), respectively.

what drop until *d* reduces down to ~20 nm, as shown in **Figure 6(a)**. In this strengthening regime, σ_y obeys the empirical H-P relationship, i.e., $\sigma_y \propto d^{-0.5}$, at $d \geq$ ~800 nm. Below this grain size, the strength of UFG/NC Cu thin films can be well captured by Eqs. (1) and (2). Also, **Figure 6(b)** clearly shows σ_y increases with decreasing *h* down to nanoregime and appears to drop slightly between 20 and 50 nm thickness. In general, *d* trends to scale with *h*. Thus, the strengthening of Cu thin films results from the constraints of both *d* and *h* on dislocation nucleation and motion.

The attainment of both strength and ductility is a vital requirement for most structural materials; unfortunately these properties are generally mutually exclusive. This general belief holds true for these NS metallic thin films/foils, such as Cu and Ni. For example, Niu et al. [23] studied the tensile ductility of NC Cu thin films with thickness spanning from 60 to 700 nm by characterizing the critical strain to nucleate microcracks, and revealed the fashion of "smaller is stronger and smaller is less ductile." The limited tensile ductility in NS thin films can be ascribed to the lack of strain hardening and grain geometry. In particular, the NC thin films with columnar grains are more favorable to exhibit quite limit uniform tensile elongation, because the insufficient room in NC grains does not permit involving intragranular dislocation interaction and entanglement and cracks are easier to propagate along columnar GBs [58]. This intrinsic limitation promotes plastic instabilities such as necking or cracking.

By far, three available strategies are presented that demonstrate enhancement of ductility in NC metals, including engineering grain-size distributions [59], embedding growth nanotwins [21], and designing high twinnability NC metals [60]. Gianola et al. [6] has uncovered that the stress-assisted grain growth has a dynamic effect on the macroscopic mechanical properties of free-standing NC Al thin films; extended ductility can be realized along with a concurrent loss in strength in comparison to tests in which no grain growth was observed. Therefore, this twinning-mediated grain growth mechanism unveiled in NT Ni [37, 47] seems to synergically combine the merits of (deformation/growth) nanotwins and grain growth mentioned above, being a novel and promising method to enhance the tensile ductility of NS metals for their performance optimization.

3.3.2. Strain-rate sensitivity

The plastic deformation kinetics in NS metals could be investigated to shed light on the strength-ductility tradeoff. It is well known that a material's strain rate dependence is usually quantified through the power law relationship: $\sigma = \sigma_0 \dot{\varepsilon}^m$ [61]. The strain-rate sensitivity (SRS) of a material can be characterized by two key kinetic signatures of deformation mechanisms, i.e., SRS index (m) and activation volume (V^*), both of which correlated via the expression $m = \frac{\partial \ln(\sigma)}{\partial \ln(\dot{\varepsilon})} = \frac{\sqrt{3}\ k_B T}{\sigma V^*}$. The former characterizes the rate-controlling process, while the latter characterizes deformation kinetics in a metal. From a series of experiments, the SRS m is summarized **Figure 7(a)** and **(b)** for UFG/NC nontwinned [62–69] and NT [37, 70, 71] FCC metals (i.e., Cu and Ni). **Figure 7(a)** shows that m for several typical small-scaled Cu materials (e.g., NC Cu, single and multicrystalline Cu micropillars) increases monotonically with decreasing their *size parameters*, with $m > 0.01$ for small-scaled NS Cu. This trend is similar to the d-effect in other nontwinned FCC metals, such as Ni [64–67], and to the λ-effect in NT Ni foils [37, 71] and bulk NT Cu [70] that are presented in **Figure 7(b)**. The large m achieved in both cases can facilitate suppressing localization at high deformation rates. In contrast, the BCC metals in general exhibit the reduced m with decreasing d [62, 72]. The fundamental difference between FCC and BCC metals can be attributed to their different dislocation core structures.

Insight into the dominant deformation mechanism is often interpreted in terms of the values of activation volume V^* for plastic deformation. In CG FCC metals, a typical rate-determining process, such as the intersection of forest dislocations, gives a large V^* of the order of several hundred to a few thousand b^3 [61, 70]. At another extreme, GB sliding or GB diffusion mediated creep (Coble creep) gives a small V^* of less than $1b^3$. When the V^* is between $1b^3$ and $100b^3$, the rate process typically involves cross-slip or dislocation nucleation from boundaries [70]. Recent findings have shown that there is linear relationship between the activation volume V^* and *size parameters*, such as grain size d, twin thickness λ and pillar diameter φ in the log-log plots [68, 69].

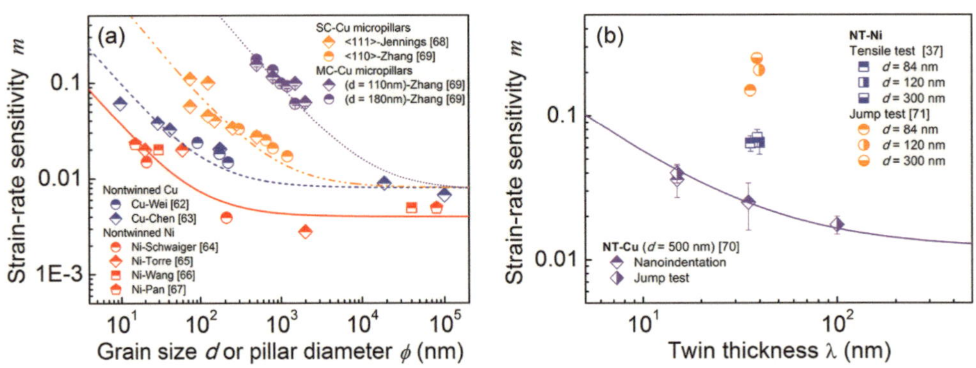

Figure 7. Strain-rate sensitivity of FCC Ni and Cu metals as a function of (a) grain size (d) or pillar diameter (ϕ) and (b) twin thickness (λ), summarized from available literatures [37, 62–71]. All the curves are visual guides.

3.3.3. Mechanical fatigue lifetime

The continuing trend of miniaturizing materials in micro- and nanodevices has led to a strong demand for understanding the complex fatigue properties of NS thin films to tailor their internal features to guarantee their reliability. Zhang and coworkers [57] investigated the fatigue behavior of NC Cu thin films with thickness spanning from 60 to 700 nm on compliant substrates by *in situ* measure the change of electrical resistance with the number of cyclic loading, by adopting the method proposed by Sun et al. [24]. **Figure 8(a)** clearly shows the dependence of fatigue lifetime (N_f) of NC Cu films on h at different strain ranges ($\Delta\varepsilon$). It is found that there is a maximum N_f at the critical thickness of h = 100 nm, above which N_f monotonically increases with reducing h at a constant $\Delta\varepsilon$. While below this critical thickness,

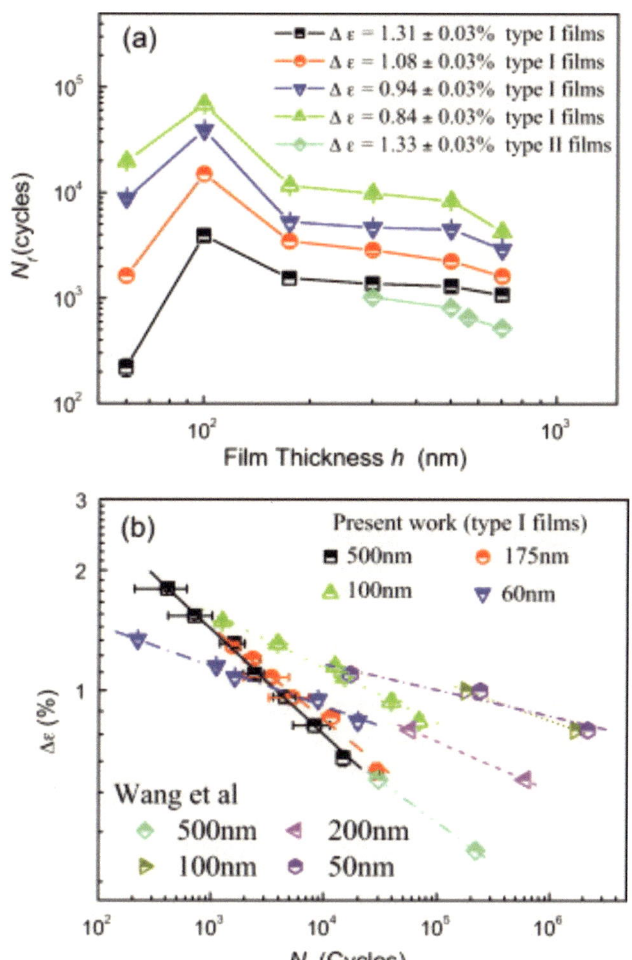

Figure 8. (a) Dependence of fatigue lifetime N_f on strain range $\Delta\varepsilon$ as a function of film thickness h for Cu thin films, respectively. (b) A comparison of the relationship of $\Delta\varepsilon - N_f$ in Cu thin films with different thickness h. Figure is taken with permission from Ref. [57].

N_f decreases with further reducing h. This is caused by the good combination of high strength (~1050 MPa) and suitable ductility (~5.5%). Luo et al. [48] recently also pointed out that in addition to the potential contribution from the high strength of nanograins (of Au), notable improvement in fatigue properties may be closely associated with twinning-mediated grain growth. For a given h, a higher $\Delta\varepsilon$ leads to a smaller N_f of the Cu thin film. Moreover, all Cu thin films exhibit the dependence of N_f on $\Delta\varepsilon$ that could be well described by the well-known Coffin–Manson relationship: $(\Delta\varepsilon/2) = \varepsilon_f(2N_f)^C$, where ε_f and C are the fatigue ductility coefficient and exponent, respectively, as shown in **Figure 8(b)**. Accordingly, with reduction in h from 700 nm with $d = 220$ nm to $h = 60$ nm with $d = 20$ nm, the surface damage morphologies change from extrusion/intrusion to intergranular cracks, due to the transition of deformation mechanism from dislocation-based to GB-mediated. In other words, with decreasing *size parameters* the localized accumulation of plastic strains within grains is hindered and the GBs take over as the preferred site for damage formation, implying the availability and activation of bulk dislocation sources become more limited in NC metals. This is consistent with the postmortem TEM observations by Zhang et al. [73].

4. Grain boundary segregation in nanocrystructured metallic materials

During the past two decades, NS metallic materials have received considerable attention owing to their unique, often desirable properties for engineering applications, whereas they manifest two adverse properties: low ductility and microstructural instability as mentioned earlier. This is because the high energy GBs associated with high mobility can absorb abundant dislocations, resulting in low dislocation storage inside grains [2, 3]. Therefore, a universal strategy to remarkably enhance/improve the mechanical properties and thermal stability of these NS materials is to manipulate their multihierarchical microstructures by embedding atoms/clusters or nanoparticles in grain interiors to increase dislocations storage and at GBs to prevent grain growth by reducing GB mobility [74]. Fortunately, alloying opens an available avenue to achieve such an idea about microstructure-sensitive design to improve materials' properties by tuning solute distributions, in particular, GB segregation, in NS thin films to achieve thermodynamically stable or metastable states [75–81]. The addition of an alloying element has fundamental thermodynamic implications for NC metals, which can explain the unique ability of alloyed systems to exhibit fine-grained structures [8–13, 75–81]. Specifically, Schuh's group [11–13] recently developed a theoretical framework for a regular NC solution (RNS) that incorporates GB segregation and further built an insightful nanostructure stability map for design alloys with positive enthalpy.

In what follows, we mainly address alloying effects on microstructural evolution on the one hand, and on the mechanical properties on the other in three categories of typical binary Cu-based film systems, i.e., Cu-Zr, Cu-Al, and Cu-W. This division of three typical binary systems is based on the consideration of mixing enthalpy (H_{mix}) and the conventional bulk binary diagram under equilibrium states (at RT), and can be extended to other systems like Ni-based binary alloy.

http://dx.doi.org/10.5772/66598

4.1. Alloying effects on microstructure and mechanical properties in the Cu-Zr model system

In such system that has a very negative enthalpy of mixing, only elemental Cu and intermetallic Cu-Zr phases coexist at room temperature under equilibrium state. However, nonequilibrium MS can result in the coexistence of solute (Zr) atoms/clusters, Cu-Zr intermetallic particles, and Cu-Zr amorphous phase in the as-deposited alloyed thin films to achieve multihierarchical microstructures, thereby facilitating the combination of high strength and ductility.

Zhang et al. [82] systematically investigated the microstructural evolution, mechanical properties, and deformation mechanisms of NS Cu thin films alloyed with Zr. It is found that Zr addition significantly changes the microstructures of NS Cu thin films. A strong (100) texture observed in the pure Cu film is strongly suppressed while the (110) texture is favorably promoted in the Cu-0.5 at.% Zr and Cu-2.0 at.% Zr films. When the Zr content is up to 8.0 at.%, the (100) and (110) peaks disappear and the (111) peak is also highly weakened, associated with an obvious amorphization tendency. The underlying reason for the change of crystallographic orientations of Cu-Zr alloyed thin films can be attributed to the effect of reduced GB energy caused by GB segregation on the competition between surface energy and strain energy [82].

Along with the crystallographic orientations change, the GB microstructures of Cu-Zr alloyed thin films also change with Zr doping, as displayed in **Figure 9**. Zhang et al. [82] uncovered that in the Cu-0.5 at.% Zr film, some nanosized $Cu_{10}Zr_7$ precipitates occasionally observed at the GBs, as indicated in **Figure 9(a)** and **(b)**, associated with notable GBs segregation of Zr,

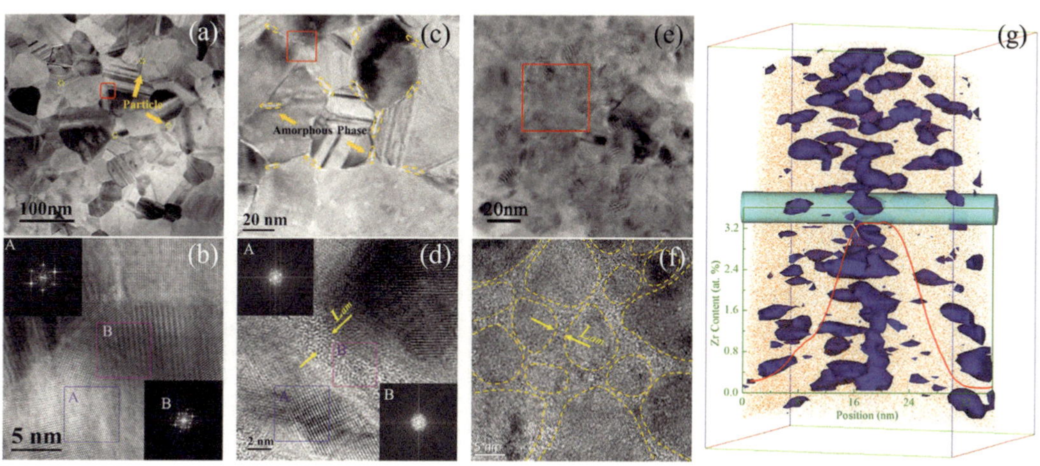

Figure 9. Representative TEM and HRTEM images demonstrating the architectured microstructures in the Cu-0.5 at.% Zr (a, b), Cu-2.0 at.% Zr (c, d), and Cu-4.0 at.% Zr (e, f) films. (g) The 3DAP image of the Zr segregation at the GB and the variation of concentration of Zr at different position along the line. Figure subparts (a–f) are taken with permission from Ref. [82].

see **Figure 9(g)**. Actually, besides having an important role in reducing GB energy, GB segregation can drive the formation of new interfacial structures at the GBs. In the Cu-2.0 at.% Zr film, discontinuous amorphous phases are frequently observed at the GBs, as indicated in **Figure 9(c)** and **(d)**. When the Zr addition is up to 4.0 at.%, GBs are unclear and continuous amorphous phase is distributed along the GBs, as shown in **Figure 9(e)** and **(f)**. Their TEM findings are consistent well with the XRD results mentioned above that Zr addition in host metal of Cu tends to induce amorphization.

Apart from the amorphization tendency and grain refinement, another significant change in microstructure caused by the Zr addition is the twinnability in the Cu films. Somewhat soluble Zr atoms reduce the SFE and thus increase twinning propensity, while excessive Zr addition induces sharply reduced twinning propensity. The dependence of twinnability on Zr addition was rationalized from the mechanisms of annealing twins by these authors [82], including (i) the successive and random emission of Shockley partials from GBs, and (ii) the GB migration mechanism accompanied with twins formation. However, the twin thickness monotonically decreases with increasing Zr contents in a fashion as same as the grain size. Furthermore, Zhang et al. [82] unambiguously demonstrated that the architectured microstructures, in particular, the GB complexions, significantly influence the mechanical properties, such as strength/hardness, ductility, and fatigue lifetime of NS materials, addressed below.

The most striking finding in their experiments [82] is that Zr addition offers exceptionally high values of both strength and ductility for the NS Cu thin films and both the strength/hardness and tensile ductility reach peak values at 0.5 at.% Zr addition, as shown in **Figure 10**. With further increasing Zr contents, the hardness shows slow reduction whereas the ductility exhibits sharp reduction. The high strength stems from various contributors, including solid solution (clusters) strengthening [83], GB solute segregation [84], Zener drag effect [85, 86], and GB/TB strengthening [3], in addition to the contribution from amorphous phase in high Zr contents samples [87–90]. The remarkable enhancement in ductility of the Cu-0.5 at.% Zr film stems from the stress-driven grain growth via twinning mechanism, displayed in **Figure 11**, like that in the pure ED Ni foils [37, 71] mentioned in Section 3. This is an indirect effect of Zr doping that benefits the emergence of (110)-oriented grains, leading to random crystallographic orientations, i.e., coexistence of (111), (100), and (110) grains, whose cooperative interaction is known to facilitate grain coarsening. This new finding in Cu-0.5 at.% Zr thin film challenges the conventional wisdom that improving the strength of a metal alloy is always a tradeoff that results in a loss of ductility—the property that allows a metal to deform without fracture.

Also, Zhang and his colleagues [82] explored the mechanical fatigue properties of these deposited Cu-Zr alloyed thin films. The NS Cu-Zr thin films were cyclically strained under different total strain ranges and the strain range $\Delta\varepsilon$ versus lifetime N_f curves were experimentally determined for the Cu films with different Zr addition, as shown in **Figure 12**. All these films exhibit the dependence of N_f on $\Delta\varepsilon$ that could be well described by the Coffin-Manson relationship: $(\Delta\varepsilon/2) = \varepsilon_f(2N_f)^C$, where ε_f and C are the fatigue ductility coefficient and exponent, respectively. At the strain range of 0.5–3.0, the Cu-0.5 at.% Zr film always displays

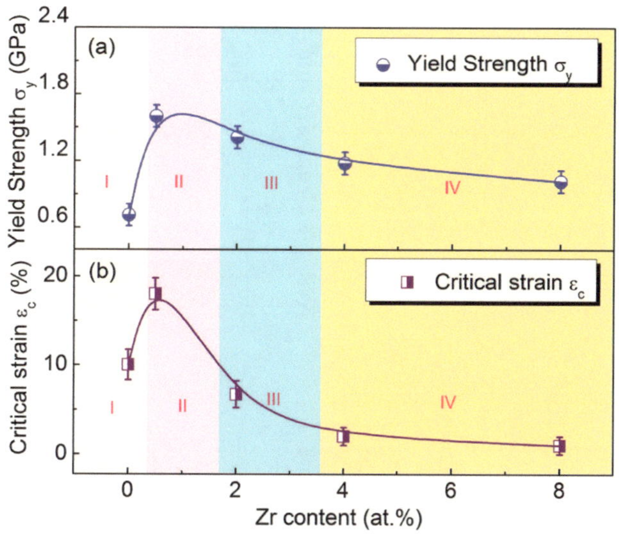

Figure 10. Dependence of yield strength ($3\sigma_y$ = Hardness) (a) and ductility characterized by the critical strain to nucleate microcracks (ε_C) (b) on Zr content. Four regimes (I, II, III, and IV) are divided which correspond to four different microstructures in the Cu, Cu-0.5 at.% Zr, Cu-2.0 at.% Zr, and Cu-Zr (Zr > 4.0 at.%) films, respectively. Figure is taken with permission from Ref. [82].

the greatest fatigue lifetime among all the films while the pure Cu film manifest the shortest one. The fatigue resistance of NS Cu films is notably enhanced via either GB segregation of Zr atoms/precipitates or amorphous phase formation of Cu-Zr, resulting in retardation of fatigue damages. Different from the uniaxial tension test, the amorphous phase plays a crucial role in the prolonged fatigue lifetime through depressing microcrack nucleation and through the arrest of intergranular cracks. Although previously atomistic simulations [87] showed that the nanoscale amorphous intergranular phase as a structure feature plays a critical role in toughening NS materials, as verified in crystalline/amorphous Cu/Cu-Zr nanolaminates [88–90], the presence of amorphous phase in Cu-Zr alloyed thin films apparently deteriorate their tensile ductility, as least at high Zr additions. This GB complexion effect on the mechanical properties is quite interesting and requires further work to reconcile the current discrepancy. Still, postmortem TEM observations in the fatigued Cu-Zr thin films verified that the occurrence of stress-driven grain growth under fatigue conditions, as same as that in tensile deformation. These findings show that the fatigue of thin metallic films remains a very attractive field of research due to the possible complex interplay of the possible deformation and fracture mechanisms.

4.2. Alloying effects on microstructure and mechanical properties in the Cu-Al model system

In this miscible system with H_{mix} close to zero, solute (Al) has a significant solid solubility in the host metal (Cu), and consequently, only a weak segregation tendency. The miscible solutes inside grains, in principle, allowing certain material properties to be finely

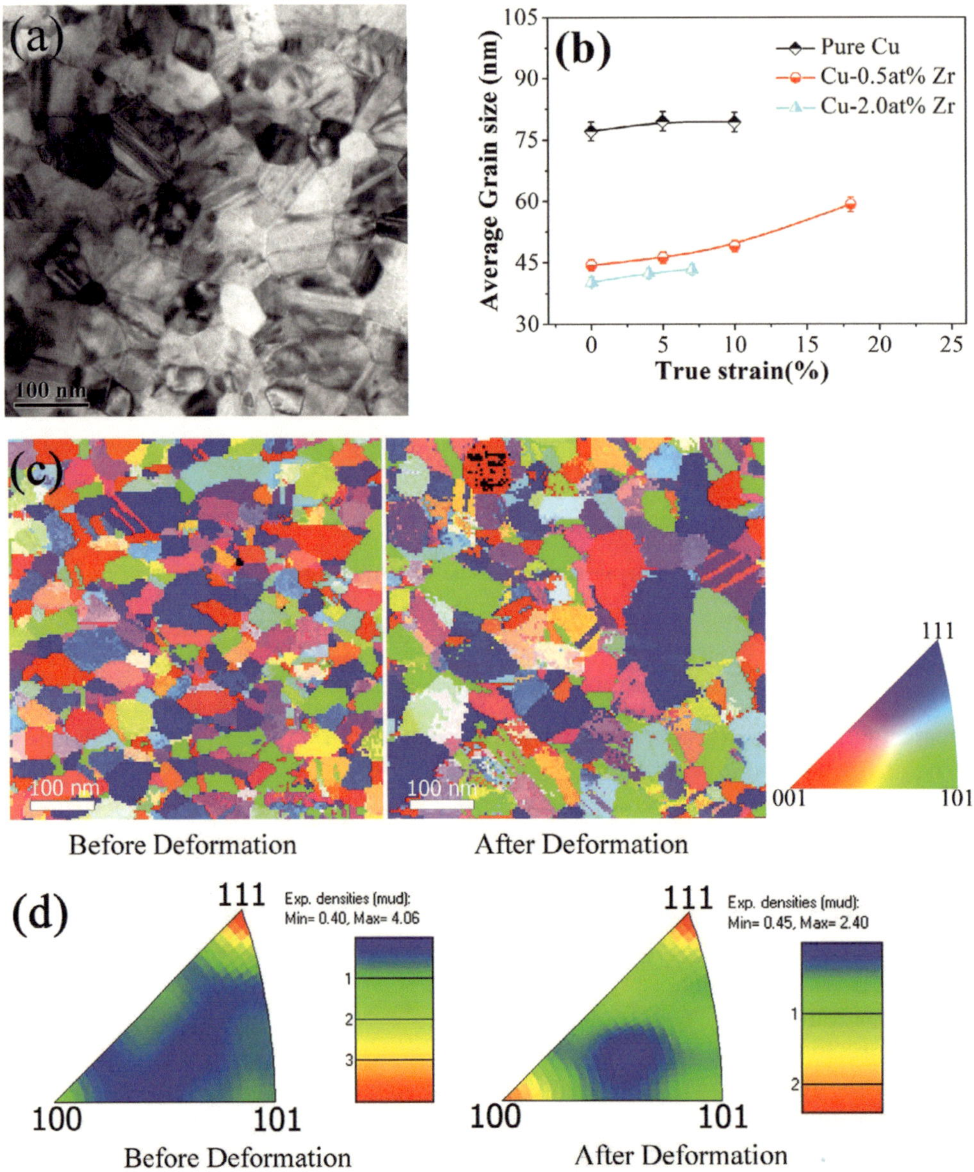

Figure 11. (a) A representative planar TEM image showing the grains in Cu–0.5 at.% Zr film stretched to 18% to demonstrate the increase in grain size. (b) Statistical results on the grain size evolution with applied strain in the pure Cu, Cu-0.5 at.% Zr and Cu-2.0 at.% Zr films. (c) Representative color-coded inverse pole figure maps from the Cu-0.5 at.% Zr film before deformation (left) and after stretching to ε_C (right), respectively, with color coding at the lower right corner. Grain growth can also be observed by comparing the images before and after deformation. (d) Corresponding orientation distributions presented as an inverse pole plot, which can be used to show the variation in orientation. Figure is taken with permission from Ref. [82].

tuned: The nucleation of partial dislocations, for instance, is stimulated by miscible solutes through markedly decreasing the SFE of the host metal [91]. For example, as NS Cu films is alloyed with Al, more nanotwins with thinner thickness are observed in the as-deposited Cu-Al thin films with lower SFE or higher fraction of Al [92, 93].

http://dx.doi.org/10.5772/66598

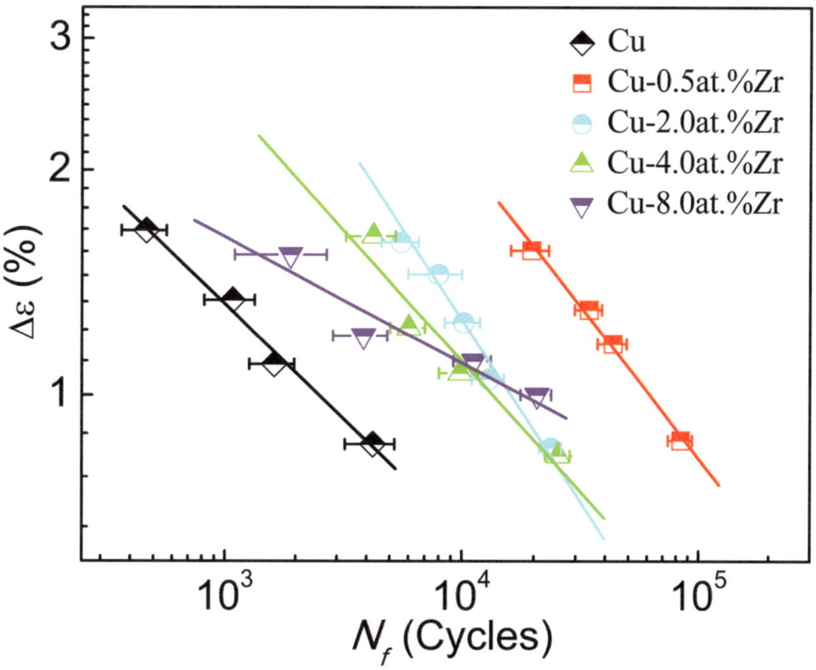

Figure 12. Dependence of the fatigue lifetime (N_f) on the strain range ($\Delta\varepsilon$) for pure Cu and Cu-Zr alloyed thin films. Figure is taken with permission from Ref. [82].

Recently, NS Cu films with different Al additions (0, 1, 5, and 10 at.%) were prepared by MS to investigate the effect of lowering SFE on microstructures and mechanical properties by Zhang et al. [92]. It is found that the Al addition motivates nanotwin formation, and promotes (111) but depresses (100) texture. With increasing Al contents, along with the refinement of grains, the morphologies of nanotwins transformed from parallel nanotwins in pure Cu to multiple nanotwins in Cu-5 at.% Al and to intersected nanotwins network in Cu-10 at.% Al, as shown in **Figure 13** as insets. Concomitantly, these Cu-Al alloyed thin films exhibit increased strength/hardness and reduced ductility with Al contents, namely, the Cu-Al films suffer from the strength-ductility tradeoff. Nevertheless, a good combination of hardness/ductility (6.2 GPa/6.3%) is achieved in the Cu-5 at.% Al film, which can be ascribed to the combined effect of texture and nanotwins [92]. At the same time, Heckman and coworkers [93] synthesized fully NT Cu-Al alloyed thin films with columnar grains and showed an increased strength of up to ~1.5 GPa that was closely related to the decrease in grain size or increase in Al content. Moreover, the ductility could be improved with decreasing the nanotwin thickness [93]. Except for the amorphous phase reinforced effect, all the strengthening mechanisms mentioned in the Cu-Zr model system play important roles in the strength of Cu-Al system. Also, Schäfer et al. [91] suggested that the details of the element distribution in the GBs are of great importance for the yield strength of the miscible alloy. The initial energetic state of the GB controls the barrier for the onset of deformation mechanisms, which is correlated to the maximum strength. Specifically, the formation of stacking faults and coherent TBs leads to material softening at high strains, because they provide additional dislocation sources. This is similar to that in pure Cu [21] and Ni [37] with very thin twins.

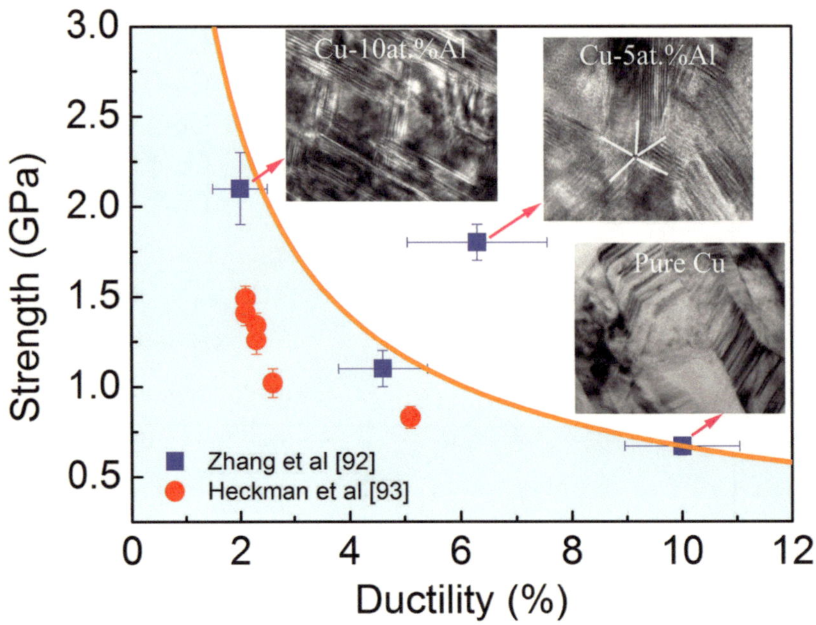

Figure 13. The strength-ductility tradeoff in Cu-Al alloyed thin films from literatures [92, 93]. Insets are the corresponding internal feature of Cu-Al films prepared by Zhang et al. [92].

In order to investigate the GB character evolution associated with the observed grain growth in this binary system, Brons and Thompson [94] carried out the *in situ* TEM observations on a sputter-deposited Cu-20 at.% Ni alloyed film that was annealed within a TEM equipped with the PED. It is found that alloying leads to a preferential evolution of particular grains. The onset of annealing resulted in multiple textures and grain growth evolution. It is clear from these results that significant differences in coincidence site lattice (CSL) boundary evolution occurred with the addition of Ni to the Cu thin film. The boundary fraction of Σ9, for the alloy film, was a factor of five larger than the pure Cu film and these boundaries were notably bounding several (but not all) of these larger grains. Additionally, Σ11 boundaries showed an increase in their fraction as compared with the elemental Cu grain growth evolution. They proposed that the Ni additions occupied subinterface sites as a result of higher surface tension of Ni which impedes other CSL boundaries motion [94]. Though solute segregation has been proposed as an efficient and effective way to stabilize NC grain structures (e.g., in Ni-W films [83]), solute preference to specific CSL boundaries can result in abnormalities in grain growth and lead to destabilization of the grain structure (e.g., in Cu-Ni films [94]). These results suggest that the GB segregation is much more complex. However, most previous numerical studies arbitrarily place solute atoms at GBs to determine their effect on GB energy [95–98], only some limited studies have been conducted to study the correlation between the CSL boundaries and solute segregation [94, 99, 100]. Therefore, further works are urgently needed to focus on how GB character is influenced with solute alloying.

4.3. Alloying effects on microstructure and mechanical properties in the Cu-W model system

It is well recognized that the Cu-W system is an essentially immiscible one characterized with a quite positive H_{mix} of about +22 kJ/mol. There does not exist any Cu-W compound in its equilibrium phase diagram. Actually, for the most studied immiscible binary alloyed films (such as Cu-W [101], Cu-Cr [102], and Cu-Ta [103]), nonequilibrium NS alloys, i.e., supersaturated solid solutions (SSSs), can be obtained by MS. Compared with the elemental (Cu) thin films, it is normally expected that the marked solute segregation that alters GB characters could occur and the solute drag effect could yield smaller grains [8, 12], both of which affect the propensity of nanotwin formation in binary Cu-based thin films.

Vüllers and Spolenak [101] recently prepared the "immiscible" Cu-W thin films with different W contents on silicon substrates using MS, and clearly demonstrated that these NC Cu-W thin films transit from the SSSs in a metastable as-deposited state to fully phase separated interpenetrating networks after annealing at 750°C, as shown in **Figure 14**. The W additions notably change the microstructural configurations of crystalline Cu thin film that has a distinctive columnar superstructure consisting of large numbers of partially even equiaxed grains and occasionally occurring twins, in the as-deposited state. While the columnarity dominant for pure Cu is still present in a 5 at.% W film, the subordinate structure making up the single columns in the pure Cu cross-section cannot be observed any longer. Films of higher W contents up to and around 30 at.% W do not exhibit a distinct crystalline structure in the as-deposited state. Subsequently, they measured the hardness and modulus of these Cu-W thin films as function of the W content at different states. It can be deduced that W content strongly influences the film's mechanical performance. As a whole, both hardness and modulus increase with increasing W contents, as shown in **Figure 14**. However, they did not perform quantitatively calculation of the strength of Cu-W thin films. Harzer et al. [102] quantitatively evaluated the hardness of metastable Cu-Cr alloyed thin films which are stable below ~170°C, and further correlated it with respect to film compositions and grain sizes in

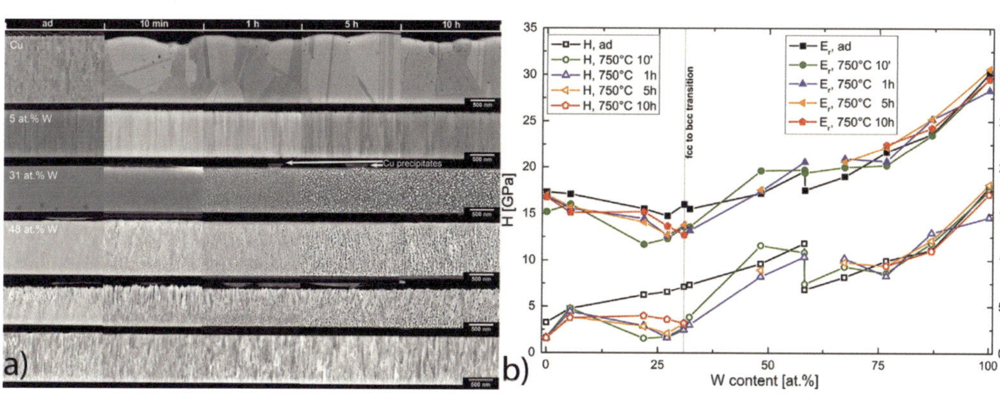

Figure 14. (a) Cross-sectional SEM images (BSE) of as-deposited and annealed Cu-W thin films (left). (b) Hardness and reduced Young's modulus as function of compositional fraction of W with standard deviation error bars for the as-deposited and annealed states (right). Figure is taken with permission from Ref. [101].

terms of several strengthening mechanisms. They concluded that the hardening of the Cu-Cr films is mainly caused by grain size refinement whereas the effects of solid solution hardening can be neglected. Nevertheless, they did not consider the contributions from the global effect of solute atoms on the matrix [83] and GB segregation [84] to the measured hardness.

Numerous atomistic simulations have demonstrated that GB segregation can remarkably stabilize the grains and enhance the strength/hardness of alloyed systems, such as Cu-Ta [15] and Cu-Nb [84]. Using molecular dynamics simulations with an angular-dependent interatomic potential, Frolov et al. [15] investigated the Ta doping effect on the barrier for grain coarsening and robust performance of NC Cu-6.5 at.% Ta alloys. It is found that Ta segregation at GBs notably increases structural stability and mechanical strength, compared with their siblings with a uniform distribution of the same amount of Ta. With increasing temperature, the Ta atoms agglomerate and segregate at GBs in the form of nanoclusters. These nanoclusters effectively pin GBs and thus prevent grain growth. Vo et al. [84] also revealed that alloying additions that lower GB energy were found to dramatically increase the yield strength of the alloy, with dilute Cu–Nb alloys approaching the theoretical strength of Cu. Their findings indicate the strength is not controlled by the grain size alone, but rather by a combination of both the molar fraction of GB atoms and the degree of GB relaxation, as captured via a new strengthening model for the NC materials. Based on the finding that strength increases with

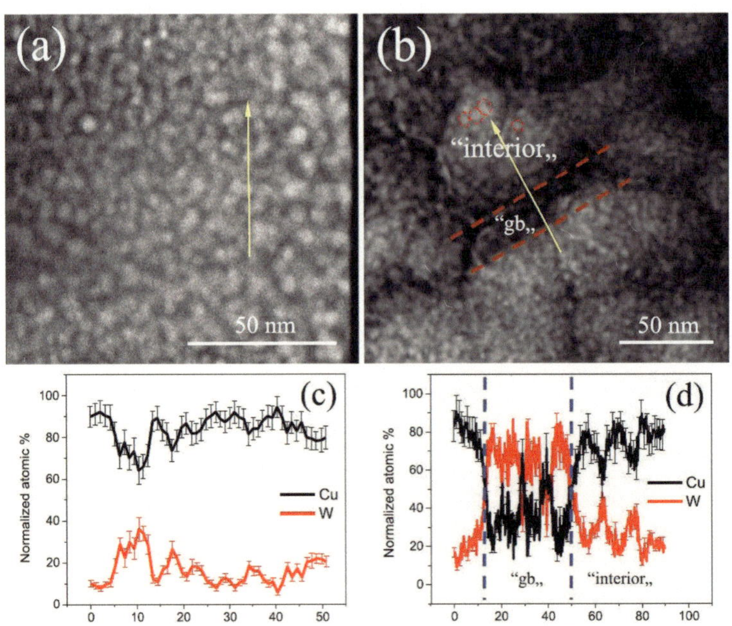

Figure 15. Cross-sectional (a) and planar-view (b) HAADF-TEM images of the annealed Cu(W)-14 thin film. The planar-view TEM image shows the prevailing columnar morphology with relatively broad grain (column) boundary regions between the columns. EDS analyses (c, d) performed by TEM reveal an inhomogeneous solute atom distribution indicating the nanoscale decomposition process. Figure is taken with permission from Ref. [104].

increasing atomic volume of the solute, they also predicted the possibility of achieving a theoretical strength in Cu by doping suitable solute atoms [84].

In parallel, Csiszár et al. [104] investigated the stability of NT Cu-W alloyed films during annealing in the range of 30–600°C, compared with their Ni-W and Ag-W NT siblings. A major, microstructural difference observed for all films upon annealing is the redistribution of the alloying element (W) content. In the case of Cu-14 at.% W, a significant redistribution of W was detected by TEM and EDS (see **Figure 15**), similar to the case of Ag-13 at.% W film but far different from that of Ni-12 at.% W film associated with a redistribution of the W atoms on an apparently very fine spatial scale. Their TEM analysis shows that an obviously nanoscale phase separation emerges throughout the Cu-14 at.% W film (see **Figure 15**). The size and the composition of the nanoinclusions at the GBs and in the grain (column) interiors are different, see **Figure 15(d)**. At the GBs, the average precipitates (rich in W) have dimensions of about 5–6 nm in diameter and in the grain interiors the precipitates (rich in Cu) are twice as large, see **Figure 15(d)**. Interestingly, the TBs are largely preserved in Ag-W and Ni-W films, whereas they completely disappear in Cu-W films. They attributed this unique phenomenon to an altered faulting energy, due to change in the amount of W segregated at TBs and to the evolution of nanosized precipitates [104]. This systematical, representative study of W-alloyed, heavily faulted NS thin films not only provides deep insights into understanding the atomic interactions in the binary alloyed films with high positive H_{mix} like Cu-W system, but also benefits us to tune their microstructural stability and mechanical properties in future.

5. Summary

The metallic thin films become essential structural materials in micro- and nanodevices and refining grain size into nanoscale indeed can notably increase their strength and strain-rate sensitivity, whereas they undergo the strength-ductility tradeoff on the one hand and suffer from unstable microstructure, i.e., grain growth, on the other. How to defect the conflict between strength and ductility and simultaneously retain highly stable microstructure is a grand challenge in the material community. The GB segregation engineering seems to open a promising avenue for the design of alloyed thin films with superior property combinations by tuning their multihierarchical structures utilizing alloying additions. The twinning-mediated grain growth is a novel and effective method to toughen the NS FCC metals and alloys with exceptionally high values of both strength and ductility. The effects of GB complexions on static and dynamic properties are far different in the alloyed thin films, and more works require to be performed in the future.

Acknowledgements

This work was supported by the National Natural Science Foundation of China (Grant Nos. 51621063, 51321003, 51571157, 51322104 and 51201123) and the 111 Project of China (B06025).

GL thanks the support from the National Science Find for Distinguished Yong Scholars. JYZ is grateful for Natural Science Basic Research Plan in Shaanxi Province of China (Program No. 2015JM5158) and China Postdoctoral Science Foundation (2016M590940) for part of financial support.

Author details

Jinyu Zhang*, Gang Liu* and Jun Sun*

*Address all correspondence to: jinyuzhang1002@mail.xjtu.edu.cn, lgsammer@mail.xjtu.edu.cn and junsun@mail.xjtu.edu.cn

State Key Laboratory for Mechanical Behavior of Materials, Xi'an Jiaotong University, Xi'an, PR China

References

[1] Arzt E. Size effects in materials due to microstructural and dimensional constraints: a comparative review. Acta Materialia. 1998;**46**(19):5611–5626.

[2] Meyers M.A., Mishra A., Benson D.J. Mechanical properties of nanocrystalline materials. Progress in Materials Science. 2006;**51**(4):427–556.

[3] Dao M., Lu L., Asaro R.J., De Hosson J.T.M., Ma E. Toward a quantitative understanding of mechanical behavior of nanocrystalline metals. Acta Materialia. 2007;**55**(12):4041–4065.

[4] Jin M., Minor A.M., Stach E.A., Morris Jr J.W. Direct observation of deformation-induced grain growth during the nanoindentation of ultrafine-grained Al at room temperature. Acta Materialia. 2004;**52**(18):5381–5387.

[5] Zhang K., Weertman J.R., Eastman J.A. Rapid stress-driven grain coarsening in nanocrystalline Cu at ambient and cryogenic temperatures. Applied Physics Letters. 2005;**87**:061921.

[6] Gianola D.S., Van Petegem S., Legros M., Brandstetter S., Van Swygenhoven H., Hemker K.J. Stress-assisted discontinuous grain growth and its effect on the deformation behavior of nanocrystalline aluminum thin films. Acta Materialia. 2006;**54**(8):2253–2263.

[7] Rupert T.J., Gianola D.S., Gan Y., Hemker K.J. Experimental observations of stress-driven grain boundary migration. Science. 2009;**326**(5960):1686–1690.

[8] Kirchheim R. Grain coarsening inhibited by solute segregation. Acta Materialia. 2002;**50**(2):413–419.

[9] Kirchheim R. Reducing grain boundary, dislocation line and vacancy formation energies by solute segregation. I. Theoretical background. Acta Materialia. 2007;**55**(15):5129–5138.

[10] Kirchheim R. Reducing grain boundary, dislocation line and vacancy formation energies by solute segregation: II. Experimental evidence and consequences. Acta Materialia. 2007;**55**(15):5139–5148.

[11] Trelewicz J.R., Schuh C.A. Grain boundary segregation and thermodynamically stable binary nanocrystalline alloys. Physical Review B. 2009;**79**(9):094112.

[12] Chookajorn T., Murdoch H.A., Schuh C.A. Design of stable nanocrystalline alloys. Science. 2012;**337**:951–954.

[13] Murdoch H.A., Schuh C.A. Estimation of grain boundary segregation enthalpy and its role in stable nanocrystalline alloy design. Journal of Materials Research. 2013;**28**(16):2154–2163.

[14] Millett P.C., Selvam R.P., Saxena A. Stabilizing nanocrystalline materials with dopants. Acta Materialia. 2007;**55**(7):2329–2336.

[15] Frolov F., Darling K.A., Kecskes L.J., Mishin Y. Stabilization and strengthening of nanocrystalline copper by alloying with tantalum. Acta Materialia. 2012;**60**(5):2158–2168.

[16] Smith D. Thin-film deposition: principles and practice. New York: McGraw Hill Professional; 1995.

[17] Geng H. Semiconductor manufacturing handbook. 1stst ed. Blacklick: McGraw-Hill Professional; 2005.

[18] Erb U., Aust K.T., Palumbo G. Electrodeposited nanocrystalline metals, alloys and composites. In: Koch C.C., editor. Nanostructured materials – processing, properties, and applications. 2nd ed. New York: William Andrew; 2007. pp. 235–292.

[19] Erb U. Electrodeposited nanocrystals: synthesis, properties and industrial applications. Nanostructured Materials. 1995;**6**(5–8):533–538.

[20] Guo N.N., Zhang J.Y., Cheng P.M., Liu G., Sun J. Room temperature creep behavior of free-standing Cu films with bimodal grain size distribution. Scripta Materialia. 2013;**68**(11):849–852.

[21] Lu L., Chen X., Huang X., Lu K. Revealing the maximum strength in nanotwinned copper. Science. 2009;**323**(5914):607–610.

[22] Lu L., Shen Y.F., Chen X., Qian L., Lu K. Ultrahigh strength and high electrical conductivity in copper. Science. 2004;**304**(5669):422–426.

[23] Niu R.M., Liu G., Wang C., Zhang G.J., Ding X.D., Sun J. Thickness dependent critical strain in submicron Cu films adherent to polymer substrate. Applied Physics Letters. 2007;**90**(16):161907.

[24] Sun X.J., Wang C., Zhang J., Liu G., Zhang G.J., Ding X.D., et al. Thickness dependent fatigue life at microcrack nucleation for metal thin films on flexible substrates. Journal of Physics D: Applied Physics. 2008;**41**:195404.

[25] Yamakov V., Wolf D., Phillpot S.R., Mukherjee A.K., Gleiter H. Deformation-mechanism map for nanocrystalline metals by molecular-dynamics simulation. Nature Materials. 2004;**3**:43–47 .

[26] Yamakov V., Wolf D., Phillpot S.R., Mukherjee A.K., Gleiter H. Dislocation processes in the deformation of nanocrystalline aluminium by molecular-dynamics simulation. Nature Materials. 2002;**1**:45–49.

[27] Van Swygenhoven H., Derlet P.M., Frøseth A.G. Nucleation and propagation of dislocations in nanocrystalline fcc metals. Acta Materialia. 2006;**54**(7):1975–1983.

[28] Budrovic Z., Van Swygenhoven H., Derlet P.M., Van Petegem S., Schmitt, B. Plastic deformation with reversible peak broadening in nanocrystalline nickel. Science. 2004;**304**:273–276.

[29] Oh S.H., Legros M., Kiener D., Gruber P., Dehm G. In situ TEM straining of single crystal Au films on polyimide: Change of deformation mechanisms at the nanoscale. Acta Materialia. 2007;**55**(16):5558–5571.

[30] Wu X.L., Zhu Y.T. Inverse grain-size effect on twinning in nanocrystalline Ni. Physical Review Letters. 2008;**101**:025503.

[31] Zhang J.Y., Liu G., Wang R.H., Li J., Ma E., Sun J. Double-inverse grain size dependence of deformation twinning in nanocrystalline Cu. Physical Review B. 2010;**81**(17):172104.

[32] Zhang J.Y., Zhang P., Wang R.H., Liu G., Zhang G.J., Sun J. Grain-size-dependent zero-strain mechanism for twinning in copper. Physical Review B. 2012;**86**(6):064110.

[33] Mompiou F., Caillard D., Legros M., Mughrabi H. In situ TEM observations of reverse dislocation motion upon unloading in tensile-deformed UFG aluminium. Acta Materialia. 2012;**60**(8):3402–3414.

[34] Hu J.J., Zhang J.Y., Jiang Z.H., Ding X.D., Zhang Y.S., Han S., et al. Plastic deformation behavior during unloading in compressive cyclic test of nanocrystalline copper. Materials Science and Engineering: A. 2016;**651**:999–1009.

[35] Asaro R.A., Suresh S. Mechanistic models for the activation volume and rate sensitivity in metals with nanocrystalline grains and nano-scale twins. Acta Materialia. 2005;**53**(12):3369–3382.

[36] Sedlmayr A., Bitzek E., Gianola D.S., Richter G., Mönig R., Kraft O. Existence of two twinning-mediated plastic deformation modes in Au nanowhiskers. Acta Materialia. 2012;**60**(9):3985–3993.

[37] Li J., Zhang J.Y., Jiang L., Zhang P., Wu K., Liu G., et al. Twinning/detwinning-mediated grain growth and mechanical properties of free-standing nanotwinned Ni foils: grain size and strain rate effects. Materials Science and Engineering: A. 2015;**628**:62–74.

[38] Schiøtz J., Jacobsen K. W. A maximum in the strength of nanocrystalline copper. Science. 2003;**301**(5638):1357–1359.

[39] Shan Z., Stach E., Wiezorek J., Knapp J., Follstaedt D., Mao S. Grain boundary-mediated plasticity in nanocrystalline nickel. Science. 2004;**305**(5684):654–657.

[40] Schiøtz J., Di Tolla F.D., Jacobsen K.W. Softening of nanocrystalline metals at very small grain sizes. Nature. 1998;**391**(6667):561–563.

[41] Carlton C.E., Ferreira P.J. What is behind the inverse Hall-Petch effect in nanocrystalline materials? Acta Materialia. 2007;**55**(11):3749–3756.

[42] Wang Y.M., Sansoz F., Lagrange T., Ott R.T., Marian J., Barbee Jr T.W., et al. Defective twin boundaries in nanotwinned metals. Nature Materials. 2013;**12**(8):697–702.

[43] Lu N., Du K., Lu L., Ye H.Q. Transition of dislocation nucleation induced by local stress concentration in nanotwinned copper. Nature Communications. 2015;**6**:7648.

[44] Li X.Y., Wei Y.J., Lu L., Lu K., Gao H.J. Dislocation nucleation governed softening and maximum strength in nano-twinned metals. Nature. 2010;**464**:877–880.

[45] Mohamed F.A. A dislocation model for the minimum grain size obtainable by milling. Acta Materialia. 2003;**51**(14):4107–4119.

[46] Edalati K., Horita Z. High-pressure torsion of pure metals: influence of atomic bond parameters and stacking fault energy on grain size and correlation with hardness. Acta Materialia. 2011;**59**(17):6831–6836.

[47] Li J., Zhang J.Y., Liu G., Sun J. New insight into the stable grain size of nanotwinned Ni in steady-state creep: effect of the ratio of effective-to-internal stress. International Journal of Plasticity. 2016;**85**:172–189.

[48] Luo X.M., Zhu X.F., Zhang G.P. Nanotwin-assisted grain growth in nanocrystalline gold films under cyclic loading. Nature Communications. 2014;**5**:3021.

[49] Kohama K., Ito K., Matsumoto T., Shirai Y., Murakami M. Role of cu film texture in grain growth correlated with twin boundary formation. Acta Materialia. 2012;**60**(2):588–595.

[50] Zhu Y.T., Wu X., Liao X., Narayan J., Kecskes L., Mathaudhu S. Dislocation–twin interactions in nanocrystalline fcc metals. Acta Materialia. 2011;**59**(2):812–821.

[51] Cao Y., Wang Y., An X., Liao X., Kawasaki M., Ringer S., et al. Grain boundary formation by remnant dislocations from the de-twinning of thin nano-twins. Scripta Materialia. 2015;**100**:98–101.

[52] Tao N.R., Lu K. Nanoscale structural refinement via deformation twinning in face-centered cubic metals. Scripta Materialia. 2009;**60**(12):1039–1043.

[53] Kraft O., Hommel M., Arzt E. X-ray diffraction as a tool to study the mechanical behaviour of thin films. Materials Science and Engineering: A. 2000;**288**(2):209–216.

[54] Yu D.Y.W., Spaepen F. The yield strength of thin copper films on kapton. Journal of Applied Physics. 2004;**95**(6):2991–2997.

[55] Zhang G.P., Sun K.H., Zhang B., Gong J., Sun C., Wang, Z.G. Tensile and fatigue strength of ultrathin copper films. Materials Science and Engineering: A. 2008;**483–484**:387–390.

[56] Gruber P.A., Böhm J., Onuseit F., Wanner A., Spolenak R., Arzt E. Size effects on yield strength and strain hardening for ultra-thin cu films with and without passivation: a study by synchrotron and bulge test techniques. Acta Materialia. 2008;**56**(10):2318–2335.

[57] Zhang J.Y., Zhang X., Liu G., Wang R.H., Zhang G.J., Sun J. Length scale dependent yield strength and fatigue behaviour of nanocrystalline Cu thin film. Materials Science and Engineering: A. 2011;**528**(25–26):7774–7780.

[58] Zhang J.Y., Zhang P., Wang R.H., Liu G., Zhang G.J., Sun J. Enhanced mechanical properties of columnar grained-nanotwinned Cu films on compliant substrate via multilayer scheme. Materials Science and Engineering: A. 2012;**554**:116–121.

[59] Wang Y., Chen M., Zhou F., Ma E. High tensile ductility in a nanostructured metal. Nature. 2002;**419**:912–915.

[60] Wang Y.M., Ott R.T., Hamza A.V., Besser M.F., Almer J., Kramer M.J. Achieving large uniform tensile ductility in nanocrystalline metals. Physical Review Letters. 2010;**105**(21):215502.

[61] Zhu T., Li J. Ultra-strength materials. Progress in Materials Science. 2010;**55**(7):710–757.

[62] Wei Q., Cheng S., Ramesh K.T., Ma E. Effect of nanocrystalline and ultrafine grain sizes on the strain rate sensitivity and activation volume: fcc versus bcc metals. Materials Science and Engineering: A. 2004;**381**(1–2):71–79.

[63] Chen J., Lu L., Lu K. Hardness and strain rate sensitivity of nanocrystalline Cu. Scripta Materialia. 2006;**54**(11):1913–1918.

[64] Schwaiger R., Moser B., Dao M., Chollacoop N., Suresh S. Some critical experiments on the strain-rate sensitivity of nanocrystalline nickel. Acta Materialia. 2003;**51**(17):5159–5172.

[65] Torre F.D., Spatig P., Schaublin R., Victoria M. Deformation behaviour and microstructure of nanocrystalline electrodeposited and high pressure torsioned nickel. Acta Materialia. 2005;**53**(8):2337–2349.

[66] Wang Y.M., Hamza A.V., Ma E. Temperature-dependent strain rate sensitivity and activation volume of nanocrystalline Ni. Acta Materialia. 2006;**54**(10):2715–2726.

[67] Pan D., Nieh T.G., Chen M.W. Strengthening and softening of nanocrystalline nickel during multistep nanoindentation. Applied Physics Letters. 2006;**88**(16):161922.

[68] Jennings A.T., Li J., Greer J.R. Emergence of strain-rate sensitivity in Cu nanopillars: transition from dislocation multiplication to dislocation nucleation. Acta Materialia. 2011;**59**(14):5627–5637.

[69] Zhang J.Y., Liang X.Q., Zhang P., Wu K., Liu G., Sun J. Emergence of external size effects in the bulk-scale polycrystal to small-scale single-crystal transition: a maximum in the strength and strain-rate sensitivity of multicrystalline Cu micropillars. Acta Materialia. 2014;**66**:302–316.

[70] Lu L., Zhu T., Shen Y., Dao M., Lu K., Suresh S. Stress relaxation and the structure size-dependence of plastic deformation in nanotwinned copper. Acta Materialia. 2009;**57**(17):5165–5173.

[71] Li J., Zhang J.Y., Zhang P., Wu K., Liu G., Sun J. Grain size effects on microstructural stability and creep behavior of nanotwinned Ni free-standing foils at room temperature. Philosophical Magazine. 2016;**96**(29):3016–3040

[72] Niu J.J., Zhang J.Y., Liu G., Zhang P., Lei S.Y., Zhang G.J., Sun J. Size-dependent deformation mechanisms and strain-rate sensitivity in nanostructured Cu/X (X = Cr, Zr) multilayer films. Acta Materialia. 2012;**60**(9):3677–3689.

[73] Zhang G.P., Volkert C.A., Schwaiger R., Wellner P., Arzt E., Kraft O. Length-scale-controlled fatigue mechanisms in thin copper films. Acta Materialia. 2006;**54**(11):3127–3139.

[74] Liu G., Zhang G.J., Jiang F., Ding X.D., Sun Y.J., Sun J., et al. Nanostructured high-strength molybdenum alloys with unprecedented tensile ductility. Nature Materials. 2013;**12**(4):344–350.

[75] Weissmüller J. Alloy effects in nanostructures. Nanostructured Materials. 1993;**3**(1–6):261–272.

[76] Seidman D.N. Subnanoscale studies of segregation at grain boundaries: simulations and experiments. Annual Review of Materials Research. 2002;**32**(1):235–269.

[77] Liu F., Kirchheim R. Nano-scale grain growth inhibited by reducing grain boundary energy through solute segregation. Journal of Crystal Growth. 2004;**264**(1–3):385–391.

[78] Raabe D., Herbig M., Sandlöbes S., Li Y., Tytko D., Kuzmina M., et al. Grain boundary segregation engineering in metallic alloys: a pathway to the design of interfaces. Current Opinion in Solid State and Materials Science. 2014;**18**(4):253–261.

[79] Lejček P., Zheng L., Hofmann S., Šob M. Applied thermodynamics: grain boundary segregation. Entropy. 2014;**16**(3):1462–1483.

[80] Saber M., Koch C.C., Scattergood R.O. Thermodynamic grain size stabilization models: an overview. Materials Research Letters. 2015;**3**(2):65–75.

[81] Peng H., Chen Y., Liu F. Effects of alloying on nanoscale grain growth in substitutional binary alloy system: thermodynamics and kinetics. Metallurgical & Materials Transactions A. 2015;**46**(11):5431–5443.

[82] Zhang P., Zhang J.Y., Li J., Liu G., Wu K., Wang Y.Q., Sun J. Microstructural evolution, mechanical properties and deformation mechanisms of nanocrystalline Cu thin films alloyed with Zr. Acta Materialia. 2014;**76**:221–237.

[83] Rupert T.J., Trenkle J.C., Schuh, C.A. Enhanced solid solution effects on the strength of nanocrystalline alloys. Acta Materialia. 2011;**59**(4):1619–1631.

[84] Vo N.Q., Schäfer J., Averback R.S., Albe K., Ashkenazy Y., Bellon P. Reaching theoretical strengths in nanocrystalline Cu by grain boundary doping. Scripta Materialia. 2011;**65**(8):660–663.

[85] Cahn J.W. The impurity-drag effect in grain boundary motion. Acta Metallurgica. 1962; **10**(9):789–798.

[86] Schwarze C., Kamachali R.D., Steinbach I. Phase-field study of zener drag and pinning of cylindrical particles in polycrystalline materials. Acta Materialia. 2016;**106**:59–65.

[87] Pan Z.L., Rupert T.J. Amorphous intergranular films as toughening structural features. Acta Materialia. 2015;**89**:205–214.

[88] Wang Y.M., Li J., Hamza A.V., Barbee Jr T.W. Ductile crystalline-amorphous nanolaminates. Proceedings of the National Academy of Sciences of the United States of America. 2007;**104**(27):11155–11160.

[89] Zhang J.Y., Liu G., Lei S., Niu J.J., Sun J. Transition from homogeneous-like to shear-band deformation in nanolayered crystalline Cu/amorphous Cu–Zr micropillars: intrinsic vs. extrinsic size effect. Acta Materialia. 2012;**60**(20):7183–7196.

[90] Wang Y.Q., Zhang J.Y., Liang X.Q., Wu K., Liu G., Sun J. Size- and constituent-dependent deformation mechanisms and strain rate sensitivity in nanolaminated crystalline Cu/ amorphous Cu–Zr films. Acta Materialia. 2015;**95**:132–144.

[91] Schäfer J., Stukowski A., Albe K. Plastic deformation of nanocrystalline Pd–Au alloys: on the interplay of grain boundary solute segregation, fault energies and grain size. Acta Materialia. 2011;**59**(8):2957–2968.

[92] Zhang P., Zhang J.Y., Li J., Liu G., Wu K., Wang Y.Q., et al. Combined effect of texture and nanotwins on mechanical properties of the nanostructured Cu and Cu–Al films prepared by magnetron sputtering. Journal of Materials Science. 2015;**50**(4):1901–1907.

[93] Heckman N.M., Velasco L., Hodge A.M. Influence of twin thickness and grain size on the tensile behavior of fully nanotwinned CuAl alloys. Advanced Engineering Materials. 2016;**18**(6):918–922

[94] Brons J.G., Thompson G.B. Abnormalities associated with grain growth in solid solution Cu (Ni) thin films. Thin Solid Films. 2014;**558**:170–175.

[95] Millett P.C., Selvam R.P., Bansal S., Saxena A. Atomistic simulation of grain boundary energetics – effects of dopants. Acta Materialia. 2005;**53**(13):3671–3678.

[96] Millett P.C., Selvam R.P., Saxena A. Molecular dynamics simulations of grain size stabilization in nanocrystalline materials by addition of dopants. Acta Materialia. 2006;**54**(2):297–303.

[97] Detor A.J., Schuh C.A. Grain boundary segregation, chemical ordering and stability of nanocrystalline alloys: atomistic computer simulations in the Ni–W system. Acta Materialia. 2007;**55**(12):4221–4232.

[98] Chookajorn T., Schuh C.A. Thermodynamics of stable nanocrystalline alloys: a Monte Carlo analysis. Physical Review B. 2014;**89**(6):064102.

[99] Bedorf D., Mayr S.G. Grain boundary doping of Cu thin films with Bi: a route to create smooth and stable nanocrystalline surfaces. Scripta Materialia. 2007;**57**(9):853–856.

[100] Keast V.J., La Fontaine A., du Plessis J. Variability in the segregation of bismuth between grain boundaries in copper. Acta Materialia. 2007;**55**(15):5149–5155.

[101] Vüllers F.T.N., Spolenak R. From solid solutions to fully phase separated interpenetrating networks in sputter deposited "immiscible" W-Cu thin films. Acta Materialia. 2015;**99**:213–227.

[102] Harzer T.P., Djaziri S., Raghavan R., Dehm G. Nanostructure and mechanical behavior of metastable Cu–Cr thin films grown by molecular beam epitaxy. Acta Materialia. 2015;**83**:318–332.

[103] Müller C.M., Sologubenko A.S., Gerstl S.S.A., Spolenak R. On spinodal decomposition in cu–34at.% ta thin films – an atom probe tomography and transmission electron microscopy study. Acta Materialia. 2015;**89**:181–192.

[104] Csiszár G., Kurz S.J.B., Mittemeijer E.J. Stability of nanosized alloy thin films: faulting and phase separation in metastable Ni/Cu/Ag-W films. Acta Materialia. 2016;**110**:324–340.

Chapter 6

Effect of Microstructure Transformations on Fatigue Properties of Laser Beam Welded Ti-6Al-4V Butt Joints Subjected to Postweld Heat Treatment

Fedor Fomin, Volker Ventzke, Falk Dorn,
Nikita Levichev and Nikolai Kashaev

Additional information is available at the end of the chapter

http://dx.doi.org/10.5772/66178

Abstract

The effect of postweld heat treatment (PWHT) on 2.6-mm-thick Ti-6Al-4V butt joints that were welded using a continuous-wave 8-kW ytterbium fibre laser was studied in terms of the microstructure, microtexture, number of welding defects, microhardness, residual stress distribution and high cycle fatigue (HCF) properties. Five types of heat treatments in the temperature range of 540–920°C are investigated. The main reasons leading to fatigue life deterioration after the laser welding process are discussed, and possible guidelines for further improvement of the HCF behaviour by a subsequent suitable type of PWHT are provided. Low-temperature annealing ($T < 600°C$) tends to harden both the base material and the welding zone without any significant effect on the fatigue properties. Heat treatments at higher temperatures ($T > 750°C$) lead to the transformation of a strong martensitic structure in the fusion zone (FZ) into more ductile coarse lamellar, which is more beneficial for fatigue performance. A suitable type of PWHT can increase the fatigue limit of a laser-welded Ti-6Al-4V butt joint by 10%; however, a slight decrease in static strength should be considered. The effect of stress relief at elevated temperatures is studied.

Keywords: titanium alloy, Ti-6Al-4V, laser beam welding, heat treatment, microstructure, fatigue, residual stress

1. Introduction

Designed in the 1950s and initially used for compressor blades in gas turbine engines, Ti-6Al-4V titanium alloy has a high specific strength, stability at temperatures up to 400°C and good corrosion resistance. These properties have brought about an extensive use of Ti-6Al-4V in a variety of applications, such as aerospace, biomedical devices and chemical processing equipment. The aircraft industry accounts for more than 80% of this usage [1]. Airframes and aeroengine parts are the two most common applications for titanium alloys. The relatively high costs of both the raw material and the part fabrication continue to hinder the wider use of titanium in applications where weight and corrosion are not critical factors. In this context, more efficient technologies for producing complex titanium structures are required to broaden the application areas of titanium alloys and minimize the inherent cost problem. Laser beam welding (LBW) is a very promising joining technique that provides the possibility of high productivity, a single-step process and the benefit of potential weight savings compared with riveting. Compared with most structural titanium alloys, Ti-6Al-4V is considered to be highly weldable. The high energy density of the laser beam enables relatively high welding speed, which results in low heat input, a narrow heat-affected zone (HAZ) and low distortion. LBW has become increasingly competitive as a joining process over the last few decades owing to significant practical advantages over electron beam welding (EBW), requiring a high vacuum environment. However, additional measures must be taken to protect the weld zone from atmospheric contamination during LBW.

A number of researchers have investigated the influence of welding parameters on the quality and mechanical properties of the laser beam welded Ti-6Al-4V butt joints [2–5]. Generally, fusion zones exhibit higher strength and lower ductility than those of the base metal, and fracture of transverse-oriented butt welds in tensile testing usually occurs in the unaffected parent material [2, 5]. Despite the generally higher tensile strength, inferior fatigue properties of laser beam welded titanium joints are commonly observed. Because Ti-6Al-4V alloy is primarily used in fatigue-critical components, poor axial fatigue behaviour of laser beam welded Ti-6Al-4V joints is one of the main factors limiting their wide industrial application. In consideration of the above-mentioned problems, more complete understanding of the reasons and mechanisms for fatigue failure of laser beam welded Ti-6Al-4V joints is of great scientific interest.

Because the control of microstructure is the primary key for obtaining required mechanical properties in titanium alloys, postweld heat treatment (PWHT), which results in the transformation of the welding zone microstructure, is one of the options aiming to achieve improved fatigue performance of laser beam welded Ti-6Al-4V butt joints. Kabir et al. [6] investigated the effects of stress relief annealing and solution heat treatment followed by ageing on the microstructure, hardness and tensile properties of autogenously welded Ti-6Al-4V butt joints. Increased microhardness in the fusion zone after stress relief annealing was observed. The joint efficiency in terms of tensile strength was maximum for the as-welded condition and was slightly decreased by nearly 5% for solution heat-treated and aged conditions. Babu et al. [7] studied the influence of two types of heat treatment (700 and 900°C) on the fatigue properties of electron beam welded Ti-6Al-4V butt joints. The specimens annealed at lower temperatures exhibited longer fatigue lives and higher tensile strength. These results were attributed to the coarsening of the microstructure in the fusion zone after PWHT at high temperatures. Tsai et al. [8] showed that heat treatment at a temperature of 790°C for 1 h followed by air cooling can

http://dx.doi.org/10.5772/66178

noticeably improve mechanical properties and provide the electron beam welded joint with a superior combination of tensile strength and ductility. Friction stir welded Ti-6Al-4V joints subjected to various types of heat treatments were studied by Edwards et al. [9], who found that hardness and strength decreased with increasing heat treatment temperature, but ductility and fatigue performance improved. It was recommended that a high heat treatment temperature can be used to obtain the best combination of strength, ductility and fatigue performance. Thus, extensive research work has been carried out on welding of Ti-6Al-4V alloy and the subsequent PWHT. However, most of the published information is focusing on the static tensile properties. There is a lack of clarity about the influence of PWHT on the high cycle fatigue (HCF) behavior of the LBW Ti-6Al-4v butt joints. The present study was undertaken to quantitatively characterize the effect of heat treatment on the HCF performance of the laser beam welded Ti-6Al-4V butt joints and link this effect with microstructural transformations that took place during PWHT.

2. Experimental

2.1. Material

The material used in this study was Ti-6Al-4V (Grade 5, AMS 4911) alloy in the form of hot-rolled and mill-annealed sheets with a thickness of 2.6 mm supplied by VSMPO-AVISMA Corporation, Russia. The chemical composition of the base material (BM) determined by energy dispersive X-ray (EDX) spectroscopy is given in **Table 1**. The mechanical properties of the material in the final rolling direction are listed in **Table 2**. Coupons of size 190 mm × 110 mm × 2.6 mm were extracted from the as-received sheet for welding experiments. The dimensions of the coupons were sufficiently long to extract seven fatigue specimens from each after welding.

	Al	**V**	**Fe**	**Si**	**Ti**
Ti-6Al-4V	6.54	3.31	0.16	0.35	89.64

Table 1. Chemical composition (wt.%) of the Ti-6Al-4V alloy in original as-received condition determined by EDX analysis.

Modulus of elasticity, E	110.9 ± 0.5 GPa
Yield strength, $R_{p0.2}$	995.3 ± 9.5 MPa
Ultimate tensile strength, R_m	1039.2 ± 11.4 MPa
Elongation at break, A	19.2 ± 0.6%

Note: Average values with standard deviation of three tested specimens.

Table 2. Mechanical properties of the Ti-6Al-4V alloy in the final rolling direction.

2.2. Laser beam welding

The welding equipment consisted of an 8-kW continuous-wave ytterbium fibre laser YLS-8000-S2-Y12 (IPG Photonics Corporation) integrated with an IXION ULM 804 CNC-controlled universal laser machine. A collimation lens of 120 mm, a focal length of 300 mm and a process fibre with a diameter of 600 μm were employed to produce a focal spot diameter of approximately 700 μm. The centre wavelength of the fibre laser was 1070 nm. The divergence half-angle of the focused multimode beam was 30.3 mrad, and the resulting beam parameter product BPP = 11.3 mm*mrad.

Prior to welding, the faying edges of the specimens were machined, ground and then thoroughly cleaned with ethanol to remove any surface oxides and contaminants. Preliminary welding experiments with small coupons were conducted to identify the optimal combination of welding parameters and obtain a good weld quality and an appropriate weld shape. The parameters finally chosen and employed for welding of coupons are listed in **Table 3**. The specimens were fixed in an open plastic box filled with Ar to protect the weld bead from air during the LBW process. The uniform Ar flow around the weld bead was provided by the injection of the shielding gas through the porous Al plate at the bottom of the box. Kashaev et al. reported based on hot gas extraction analysis that this shielding technique was very effective [5]. The welding direction was perpendicular to the rolling direction of the material.

Laser power	5500 W
Welding speed	4.0 m/min
Focal position	-3.0 mm
Filler wire	Ti Grade 5, Ø 1.0 mm
Filler wire feed rate	3.0 m/min
Shielding	Argon, 15 l/min

Table 3. Laser beam welding process parameters.

2.3. Postweld heat treatment

Postweld heat treatment was conducted using Workhorse vacuum furnace, Centorr Vacuum Industries, USA, at a vacuum degree of 0.1 Pa. The parameters of different types of PWHT are given in **Table 4**. Annealing temperatures did not exceed the β transus, which is nearly 995°C for the Ti-6Al-4V alloy [1, 2]. During heat treatment, the welded plates were hung using molybdenum wire to prevent any contact between the specimens and furnace wall. No significant distortion was observed after heat treatment. Cooling to room temperature was performed in Ar atmosphere. All heat-treated specimens were then machined from both sides before extracting the fatigue specimens.

Stress relief annealing, SRA1	540°C/4 h/AC
Stress relief annealing, SRA2	650°C/2 h/AC
Full annealing, FA1	750°C/2 h/AC
Full annealing, FA2	850°C/1 h/AC
Recrystallization annealing, RA	920°C/45 min/AC
Duplex annealing, DA	RA + SRA1

Table 4. Conditions of postweld heat treatment.

2.4. Microstructural characterization

Transverse cross sections were cut from the stable middle region of the joint for metallographic examination and microhardness testing. After sectioning, the samples were mounted, ground and polished using an oxide polishing suspension (OPS) compound. Microstructural observations were performed using both inverted optical microscopy (OM) Leica DMI 5000M and scanning electron microscopy (SEM) JEOL JSM-6490LV. Prior to light microscopy, the specimens were etched by Kroll's reagent (3% HF, 6% HNO_3, 91% distilled water) to unveil the microstructural features. For SEM investigations, a mirror-like OPS polished surface was used. SEM microstructure observations and texture analysis of the joints were conducted using secondary electrons images and electron backscatter diffraction (EBSD). The EBSD measurements were performed for a specimen area of 135 μm × 135 μm at an acceleration voltage of 30 kV, a spot size of 4.7 nA, an emission current of 75 μA, a working distance of 13 mm, a step size of 0.3 μm and a sample tilt angle of 70°. For the orientation calculation, the generalized spherical harmonic series expansion (GSHE) method was applied based on triclinic sample symmetry. The average grain size was measured using the OIM software and the results of EBSD measurements. EDX spectroscopy was used for the local chemical composition determination. For EDX analysis, SEM was operated at an acceleration voltage of 15 kV, a working distance 10 mm and a live time 150 s. The data obtained were calculated based on the standard ZAF method of correction.

2.5. Microhardness testing

Transverse cross sections of the samples for microhardness testing were prepared in the same manner as discussed for the microstructural evaluations. The Vickers microindentation hardness test was carried out using a Zwick/ZHU0,2/Z2,5 universal hardness testing machine and testXpert software. The samples were tested with a 500-g load applied for 15 s according to ASTM E384-11 [10]. The indentation spacing was 200 μm to provide the minimum recommended distance between test points [10]. This resulted in 61 indentations for each line. To investigate thickness gradients, microhardness profiles were measured at three testing positions: radiation exposure side (RES), middle of the weld (M) and the weld root side (RS) (illustrated in **Figure 9**). The distances from RES and RS lines to the edges of the specimen were 200 μm each.

2.6. Residual stress analysis

The distribution of residual stresses in the vicinity of the welding seam was measured by the hole drilling method (HDM) combined with electronic speckle pattern interferometry (ESPI) using a PRISM system, American Stress Technologies Inc. The procedure for residual stress measurement using the HDM is described in ASTM E837 standard [11]. More details about the PRISM system and the theoretical aspects can be found elsewhere [12–14].

All measurements were performed using a high-speed air turbine at rotational speeds of 30,000 rpm, 0.04 mm/s feed rate, using a two-fluted end mill of 0.6 mm diameter. The incremental HDM technique, which involves drilling in a series of small steps in depth, was used in the present work. The holes were drilled to a 0.3 mm depth in 10 steps of 0.03 mm. The geometry of welded plates for HDM measurements and locations of the drilled holes are presented schematically in **Figure 1**. The distance between neighbouring points was provided to be at least 3 mm in accordance with ASTM E-837. The distribution of residual stresses across the welding region, that is, along the *Y* axis in **Figure 1**, was measured by drilling the holes at a number of distances from the weld centreline. At least three points represent each distance in the *Y* direction.

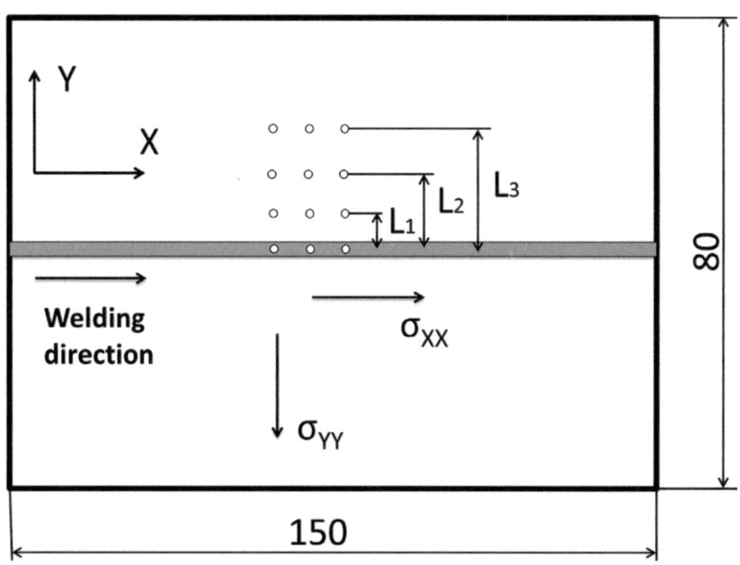

Figure 1. Geometry of welded plates and locations of drilled holes for residual stress measurements.

2.7. Fatigue testing

Load-controlled uniaxial fatigue tests were conducted at room temperature using a Testronic 100 kN RUMUL resonant testing machine. The tests were carried out in accordance with ASTM E466-07 [15] at a frequency of approximately 80 Hz and a stress ratio of $R = 0.1$. The specimen geometry with a uniform test section was chosen for fatigue testing of the LBW Ti-6Al-4V butt

http://dx.doi.org/10.5772/66178

joints in the present work. For all specimens, the welding seam was located in the centre of the gage length, and fatigue loading was applied transverse to the weld direction. The test section in the middle of the S-N specimen had a length of 20 mm (the gage length) and a width of 8 mm. Fatigue properties of the LBW Ti-6Al-4V butt joints were characterized in three conditions: as-welded, machined and heat treated + machined condition. Milling of the surfaces was conducted to remove geometrical defects after welding, such as underfills and reinforcements. The improvement of surface quality after machining was observed. The mean roughness depth, as measured using a contact profilometer in the as-received plate, was 3 μm Rz; after milling, it was 1.5 μm Rz.

For reference, base material fatigue tests were also conducted in the present study. S–N curves were obtained for the base material with the as-received surface quality and after milling. These data must be considered as well, because after machining the weldments flush with the plate surface to remove geometric imperfections, their surface quality is changed significantly, and we cannot further compare the results of the tests with the as-received base material condition. The milled base material should be considered as the reference for machined butt joints. Otherwise, the effect of removing underfills will be overestimated owing to the better quality of the specimen surface after milling. The fatigue limit was determined experimentally as the maximum stress, below which the specimen would not fracture after 10^7 cycles.

2.8. Microfractography

After fatigue testing, the ruptured specimens were selected for further fracture surface examinations. Microfractography is a useful method for understanding the relationship between fracture behaviour and microstructural characteristics. Important clues on the underlying causes of fatigue fracture may be revealed by microfractographical analysis at appropriate magnification. Fracture surfaces of the fatigue specimens were examined using OM and SEM. The SEM observations were performed in secondary electron contrast with 25 kV acceleration voltage and a working distance of 15 mm using the same machine as for metallographic analysis.

3. Results

3.1. Macroscopic appearance of the LBW Ti-6Al-4V butt joints

Visual inspection of laser weldments showed bright silver metallic surfaces from the top and root sides, indicating stable Ar shielding gas atmosphere during the LBW process. All obtained welds were fully penetrated and showed an hourglass shape. A typical transverse cross section of the obtained laser beam welded joint is shown in **Figure 2**. Three distinct regions of the weld are visible: the fusion zone (FZ), heat-affected zone HAZ and base metal (BM). The FZ was distinguished by its columnar dendrites, which grew in the direction from the fusion line to the weld centre (see **Figure 2**). It should be noted that no significant differences between the as-welded and heat-treated conditions regarding the weld geometry were found.

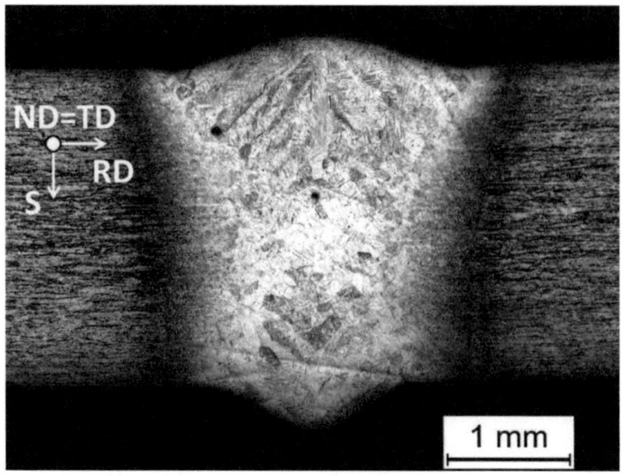

Figure 2. Transverse cross section macrograph of the laser beam welded Ti-6Al-4V butt joint.

The most frequently observed defects in laser weldments of titanium alloys are underfills and porosity [3, 5, 16, 17]. These imperfections are particularly undesirable for structures subjected to cyclic loading because they lead to stress concentration and consequently premature crack formation. The evaporation and expulsion of the molten material from the weld pool and liquid metal flow around the keyhole are dominant processes affecting the formation of underfills [2, 18], which are always present in cases of autogenous laser welding. The use of an additional filler wire in the present work allowed overfilling to be produced and geometrical weld imperfections such as underfills to be partially eliminated. However, filler wire resulted in weld reinforcements from both face and root sides. Abrupt change in the thickness due to weld reinforcement leads to stress concentration at the weld toes and roots and consequently reduces the fatigue strength of the joints. Although the macrograph presented in **Figure 2** reveals almost no underfills, a single weld cross section cannot guarantee the uniformity of the weld profile over the whole length of the seam. This problem can usually be solved by extracting more than one specimen for metallurgical examination. In the present work, the maximum measured underfill depth was approximately 70 μm, which is less than 3% of the specimen thickness. The maximum observed reinforcement was approximately 350 μm. The geometric profile imperfections of laser weldments used in the aerospace industry are strictly limited by several standards: AWS D17.1 [19] and EN 4678 [20]. In terms of underfills and weld reinforcements, EN 4678 is more stringent; the maximum allowed underfill depth for butt joints is 5% of the total thickness and maximum reinforcement is 490 μm for the specimens of the 2.6-mm-thick material. All welds in our work confidently passed the acceptance criteria in terms of the weld profile imperfections for the aerospace industry.

The spherical shape of most pores observed in the present study indicates gas-type porosity. A number of researchers have investigated the main causes of porosity when laser welding titanium alloys [18, 21, 22]. Potential sources for porosity formation are mainly from the presence of excessive hydrogen in the FZ, which is rejected upon solidification, and keyhole instability leading to the entrapment of shielding gases. The investigation of the influence of welding parameters on the amount of porosity in the laser beam welded Ti-6Al-4V butt joints can be found elsewhere [2, 3, 5].

3.2. Microstructure and microtexture

3.2.1. Base material

The as-received BM microstructure of the 2.6-mm-thick Ti-6Al-4V sheet consists of globular α grains with average grain size of 3.1 ± 0.8 μm and an intergranular retained β, as shown in **Figure 3**. This corresponds to what is known as a mill-annealed microstructure for a hot worked plate that is not fully recrystallized [1]. **Figure 3(a)** shows incompletely recrystallized regions with lamellar morphology, which are not fully transformed to equiaxed α grains upon mill annealing after a hot rolling process. The bright regions in **Figure 3(a)** are the equiaxed or lamellar α grains, and the dark regions are the intergranular β grains distributed at α grain boundaries. Texture analysis is represented by pole figures and inverse pole figures in the cross-sectional plane. It should be kept in mind that the normal to the cross section coincides with the transverse direction of the sheet, and the S direction stands for the thickness direction (see **Figure 2**). The colour in the crystal orientation map (**Figure 3(b)**) is based on a colour-coded inverse pole figure, in which different colours represent different crystallographic orientations. Large red regions in the orientation map correspond to not fully recrystallized lamellar regions. They have the same colour owing to their near-equal crystallographic orientation and were not considered for the calculation of the average grain size.

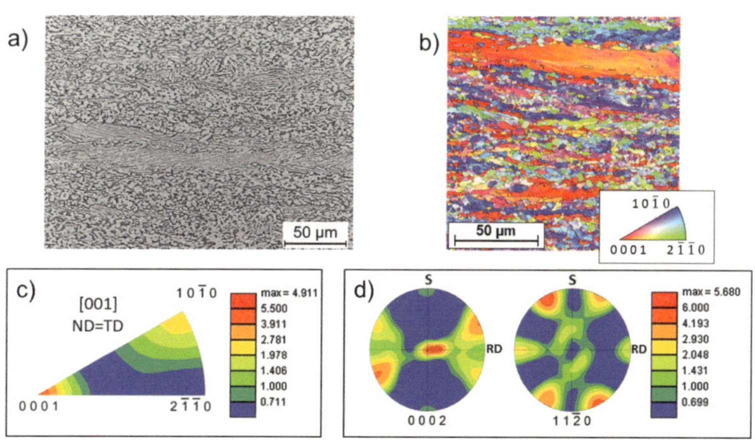

Figure 3. Microstructure of the BM in the as-received condition. (a) OM image, (b) orientation map in the cross section, (c) inverse pole figure, and (d) (0 0 0 1) and (1 1 –2 0) pole figures in the cross-sectional plane.

The base material is characterized by preferred crystal directions such as <0 0 0 1>//ND and <1 0 -1 0>//ND, as shown by the inverse pole figure in **Figure 3(c)**. The microtexture components have been determined using the orientation distribution function (ODF) at sections of $\phi_2 = 0°$ and $\phi_2 = 30°$, where ϕ_2 is an Euler angle. The results analysis showed that the Ti-6Al-4V base material contained the components (0 0 0 1)[2 -1 -1 0] ($f = 12.7$ mrd (multiple of a random distribution), (0 1 -1 0)[2 -1 -1 4] (4.3 mrd $< f <$ 9.4 mrd) and (0 1 -1 0)[2 -1 -1 0] (f = 1.6 mrd). The (0 0 0 1) pole figure shows that basal planes are aligned in both the rolling direction (RD) and transverse direction (TD) (see **Figure 3(d)**). According to the investigation of Salem [23], this type of texture indicates that the Ti-6Al-4V sheet was cross-rolled. The pole density of the (0 0 0 1) pole figure at RD shows an opening angle of 90° and corresponds to orientation bands at ($\phi_2 = 0°$, $\phi_1 = 60°$ and $\phi_1 = 120°$, $0° \leq \phi \leq 180°$) and ($\phi_2 = 30°$, $\phi_1 = 90°$, $0° \leq \phi \leq 180°$).

3.2.2. Fusion zone

The microstructure of the FZ is characterized by columnar prior β grains that grow from the HAZ in the direction opposite that of the heat flow and impinge at the weld centreline after solidification (see **Figures 2** and **4**). The FZ prior β grain size depends primarily on the weld energy input, with a higher energy input promoting a larger grain size [2, 18, 24]. In the present study, the average prior β grain size in the as-welded condition was approximately 200–300 μm. Within the prior β grains, the FZ predominantly consists of an acicular α' martensitic structure, resulting from the diffusionless $\beta \rightarrow \alpha'$ transformation upon high cooling rates encountered in the LBW process. Ahmed et al. [25] investigated the effect of different cooling rates on microstructural reactions in Ti-6Al-4V and found out that fully α' martensitic transformations take place at cooling rates above 410°C/s. The martensitic microstructure was characterized by long orthogonally oriented thin plates having acicular morphology. **Figure 4(a)** and **(b)** show a similar microstructure in the FZ of laser beam welded Ti-6Al-4V butt joints. Ahmed et al. observed preferential grain boundary formation of secondary α morphology at cooling rates in the range between 410 and 20°C/s. Because no secondary α at prior β grain boundaries was observed in the current work, according to the results of Ahmed et al., we can conclude that the cooling rate in the welding zone was high enough to provide $\beta \rightarrow \alpha'$ diffusionless transformation during LBW. The observed microstructure is typical for fusion zones of laser beam welded [2, 5] and electron beam welded [7, 8] joints.

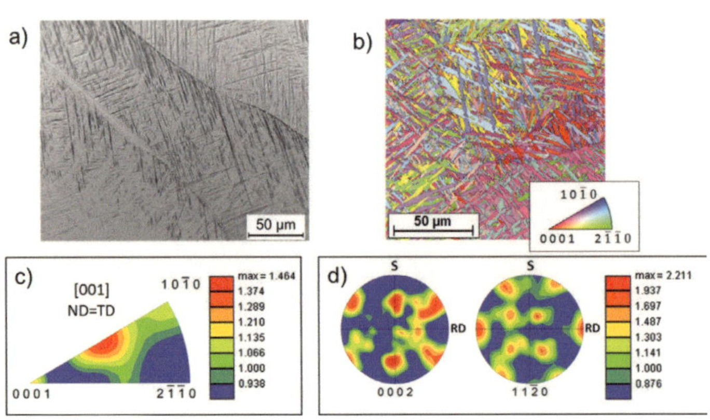

Figure 4. Microstructure of the FZ. (a) OM image of the microstructure, (b) orientation map in the cross section, (c) inverse pole figure, and (d) (0 0 0 1) and (1 1 –2 0) pole figures in the cross section plane.

Local melting and subsequent solidification of Ti-6Al-4V has changed the initial microtexture significantly. The main microtexture components in the FZ are (0 6 -6 1)[8 -3 -5 12] (f = 3.7 mrd), (0 6 -6 1)[-14 5 9 24] (f = 2.2 mrd), (0 1 -1 1)[-4 -1 5 6] (f = 3.3 mrd), (0 2 2 -3)[2 -1 -1 0] (f = 3.2 mrd) and (0 1 -1 1)[2 -1 -1 0] (f = 2.8 mrd), which were examined at 100× magnification to obtain higher statistical weight. The (0 0 0 1)[2 -1 -1 0] component (f = 2.5 mrd) is comparatively much less pronounced (**Figure 4(c)** and **(d)**) than it was in the BM (**Figure 3(d)**). The preferred direction of crystal growth was the <1 1 -2 0>//RD crystal direction during solidification. Furthermore, the presence of <1 1 -2 0>//RD fibre texture is visible.

3.2.3. Heat affected zone

HAZ displays the transition region between the acicular morphology in the FZ and globular structure in the BM. It is usually divided into two subregions based on the β transus temperature. In the HAZ adjacent to the FZ (near-HAZ), the temperatures exceed the β transus during LBW. Consequently, this region consists mostly of the transformed acicular microstructure. Because the temperatures in the HAZ adjacent to the BM (far-HAZ) were lower than the β transus, its microstructure is very similar to that of the BM. Microstructure analysis of both the near-HAZ and far-HAZ is shown in **Figure 5**.

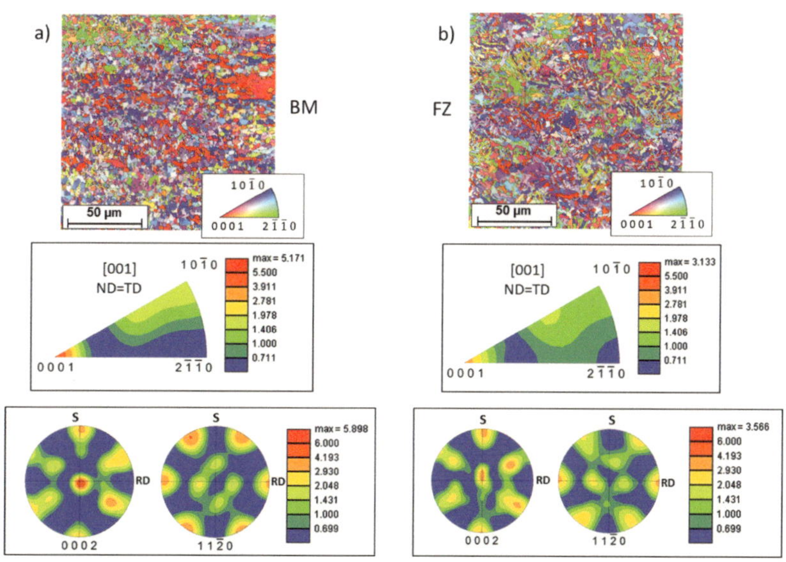

Figure 5. Microstructure of the HAZ in the as-welded condition. Orientation maps, inverse pole figures and pole figures of the HAZ adjacent to the BM (a) and the HAZ adjacent to the FZ (b).

The microstructure of the HAZ zone adjacent to the BM (1.4 mm from the weld centre) remained nearly the same after thermal cycles were imposed by the LBW process but was characterized by finer globular grains having an average grain size of 2.1 ± 0.7 μm and wider distribution of crystal directions between <0 0 0 1>//ND and <1 0 -1 0>//ND compared with that of the base material (**Figure 5(a)**). The main component is still (0 0 0 1)[2 -1 -1 0]] (12.1 mrd <

f < 12.8 mrd); however, (0 1 -1 0)[2 -1 -1 0] was superimposed by (0 8 -8 1)[2 -1 -1 0] (f = 6.5 mrd), leading to misorientations between 3.6° and 4.4°. The BM component (0 1 -1 0)[2 -1 -1 4] has been dissolved, and new components such as (0 6 -6 5)[5 -5 0 6] (f = 6.4 mrd) and (0 3 -3 -4)[14 3 -17 15] (f = 9.4 mrd) have emerged within the orientation band ($\phi_2 = 0°$, $\phi_1 = 60°$, $0° < \phi < 80°$). The misorientations of these components are 26.2° and 37.3° with regard to (0 1 -1 0)[2 -1 -1 4]. The orientation band ($\phi_2 = 0°$, $\phi_1 = 120°$, $0° < \phi < 180°$) showed the occurrence of further components such as (0 5 -5 1)[-3 1 2 5] (f = 5.6 mrd) and (0 6 -6 5)[-8 9 -1 12] (f = 4.6 mrd), resulting in misorientations of 7.7° and 24.7° with regard to (0 1 -1 0)[-2 1 1 4] (see pole figure in **Figure 5(a)**). The misorientations between the base material and heat-affected zone indicate microstructural distortion of the laser beam welded Ti-6Al-4V butt joint due to heat input and subsequent rapid cooling during the joining process.

A significant change in microstructure was observed for the near-HAZ at a distance of 1.05 mm from the FZ (**Figure 5(b)**). The equiaxed initial microstructure of the base material was transformed into an acicular morphology with a small amount of embedded globular grains. The transformation was connected with a grain refinement and a weakening of the (0 0 0 1)[2 -1 -1 0] component dominating the BM and HAZ adjacent to the BM. Some of the crystals rotated around the angles between 4.6° and 4.8°, which led to formation of (0 0 0 1)[4 -7 3 0] (5.9 mrd < f < 6.5 mrd). A further portion of crystals rotated around the <2 -1 -1 0>//RD crystal direction and tilted at an angle of 10.3°, resulting in the formation of the (0 1 -1 -1)[2 -1 -1 0] component (f = 7.3 mrd). Furthermore, the microstructure transformation led to the formation of new components such as (0 3 -3 2)[14 -12 -2 15] (f = 4.5 mrd), (0 1 -1 -1)[1 0 -1 1] (f = 2.1 mrd), (0 1 -1 -1)[-6 -1 7 8] (f = 3.9 mrd) and (0 1 -1 -1)[-17 19 -2 21] (f = 3.0 mrd) and to dissolution of (0 6 -6 5)[5 -5 0 6]. The (1 1 -2 0) pole figure shows a tendency towards the formation of <1 1 -2 0>//RD fibre texture present in the FZ also (**Figure 4(d)**).

3.2.4. Influence of heat treatment on the microstructure

EBSD analysis of heat-treated specimens revealed no significant texture transformations upon PWHT. The main texture components remained approximately the same with slight deviations in numerical values of peaks. Thus, our further attention will be focused mainly on the microstructural characteristics, which can be clearly seen from the OM observations. **Figure 6** shows the influence of PWHT on the average grain size in the BM. Heat treatments at temperatures less than 750°C did not change the grain size in the BM significantly (**Figure 7(a), (b)**). Recrystallization processes leading to coarsening of the microstructure were activated at higher temperatures starting from 800°C (FA2, DA). The maximum average grain size was achieved after DA and was 4.9 ± 1.5 μm. OM images of the BM microstructure after PWHT at temperatures higher than 750°C are shown in **Figure 7**. From this figure, it can be seen that α phase is fully recrystallized after duplex annealing, and almost all lamellar regions were transformed into equiaxed grains (**Figure 7(c)**). All other conditions of PWHT that are not shown in **Figure 7** led to almost the same microstructure as in the starting condition, and these images are omitted.

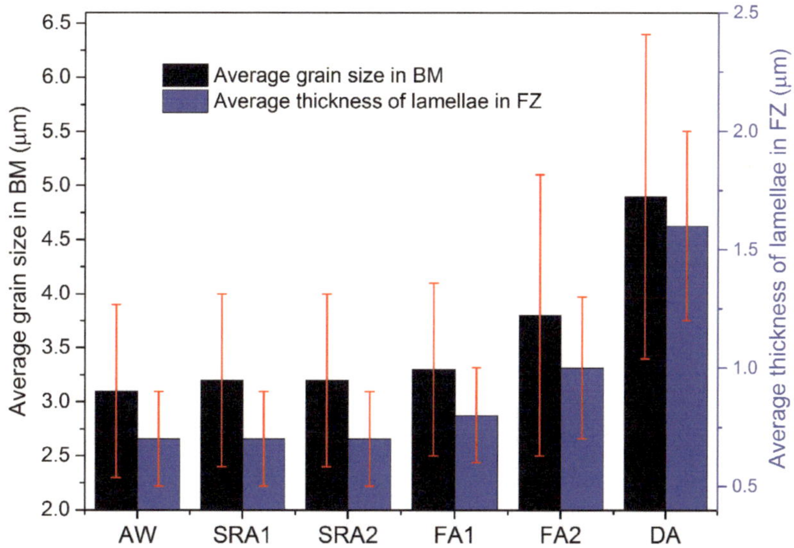

Figure 6. Influence of PWHT on the average grain size in the BM and the width of lamellae in the FZ.

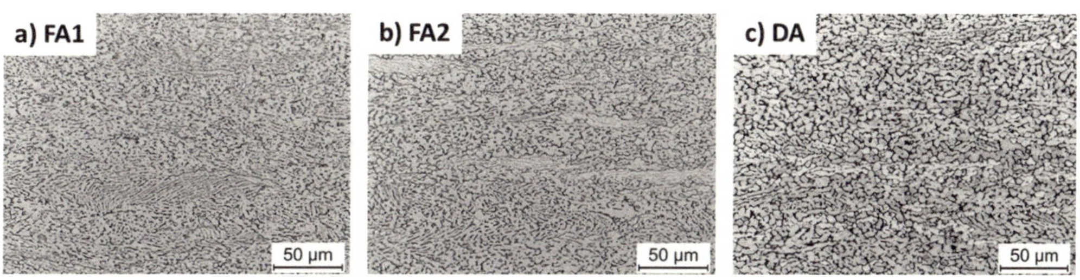

Figure 7. Influence of PWHT on the microstructure of the BM.

Light microscopic investigations yielded that heat treatments up to 650°C (SRA2) did not affect the microstructure of the FZ significantly because such low temperatures were insufficient for martensite decomposition into equilibrium $\alpha + \beta$ structure. Sallica-Leva et al. [26] studied the effects of heat treatment on the mechanical properties of an acicular α' martensite obtained by selective laser melting. In their work, the microstructure of samples heat treated at 650°C was very similar to that of the starting condition, whereas significant grain coarsening was observed after heat treatment at 800°C. The precipitation of β phase and the gradual transformation of α' into α phase by the diffusion of excess vanadium from α' to β phase were proposed to be the main events of martensite decomposition. Their results are in a good agreement with our findings. **Figure 6** shows the influence of PWHT on the average thickness of α' (α) laths in the FZ. Significant grain coarsening of the microstructure was observed after PWHT at temperatures higher than 750°C. This result is a consequence of the transformation of fine martensitic morphology into an equilibrium lamellar $\alpha + \beta$ structure and diffusion-controlled growth of obtained platelets at high temperatures (**Figure 8**). Starting from 750°C (FA1), a secondary α phase at prior β grain boundaries designated as grain boundary α [25] starts to appear (**Figure 8(b)** and **(c)**). Furthermore, the thickness of grain boundary α gradually increases with increasing PWHT temperature. After DA, the colonies of parallel α plates are formed upon recrystallization processes. In the case of the martensitic structure, fine α' needles are nearly orthogonal, and α colonies are not so pronounced (see **Figure 4(b)**). As we can see in **Figure 8(c)**, in DA condition, the α colony size is clearly visible and is approximately 20–30 μm. These α colonies should not be confused with prior β grains. The prior β grain size was not altered after PWHT; it was an order of magnitude larger than the α colony size and cannot be seen in **Figure 8**.

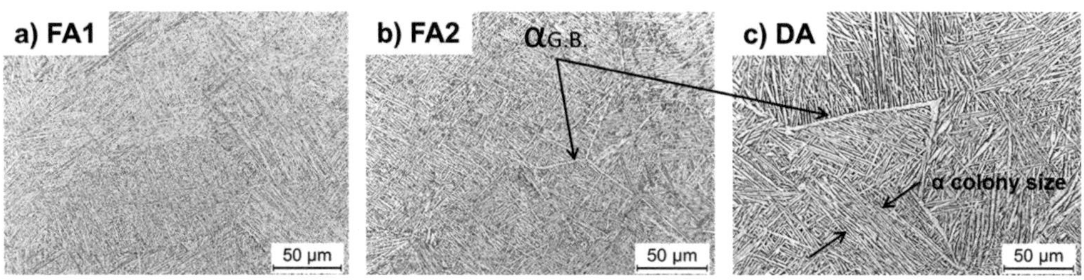

Figure 8. Influence of PWHT on the microstructure of the FZ.

Because the HAZ has a bimodal microstructure and consists of equiaxed primary α grains in transformed β matrix, the effect of PWHT on the HAZ microstructure is somehow the combination of the above-mentioned effects on the BM and FZ, that is, coarsening of primary α grains and transformation of fine martensitic structure into coarse lamellar.

3.3. Microhardness

The distribution of microhardness across the laser beam welded Ti-6AL-4V butt joint in the as-welded condition is presented in **Figure 9**. No significant difference in microhardness profiles among three testing positions was found; that is, no thickness gradient was observed in the present study. The average microhardness value of the base material was found to be 336 ± 8 HV 0.5. The FZ exhibited the highest average microhardness, approximately 396 ± 10 HV 0.5 (roughly 18% greater than that in the BM), and it decreased abruptly as the distance from the FZ line increased. The microhardness distribution within the FZ was quite uniform without significant deviations from the average value. This result is related to the use of Ti Grade 5 as a filler wire material. In the case of LBW with Ti Grade 2 (commercially pure Ti) filler material, a significant decrease in microhardness in the centre of the FZ was observed [27].

The increase in microhardness from the BM via the heat-affected zone to the FZ centre is correlated with local changes in microstructure, which were activated during LBW and subsequent cooling. The occurrence of maximum hardness in the FZ is related to the formation of a strong martensitic structure due to high cooling rates upon solidification. Acicular α' phase produced by the diffusionless transformation from the high-temperature β phase field exhibits higher strength and lower ductility, which are attributable to the fine size of martensitic plates and high defect density [28]. The influence of microtexture plays a minor role because <0 0 0 1>//ND and <1 0 -1 0>//ND crystal directions were aligned parallel to the indentation direction in both the BM and HAZ despite decreasing axial intensity. HAZ is characterized by strong inhomogeneity and plays a role of a transition zone from acicular martensitic morphology within the FZ to equiaxed microstructure in the BM. The strong spatial variation of the microstructure, namely, the decrease in the martensitic content, leads to a high gradient of microhardness inside the HAZ. Squilace et al. [2] reported that the hardness gradient in the HAZ is inversely proportional to the heat input during LBW.

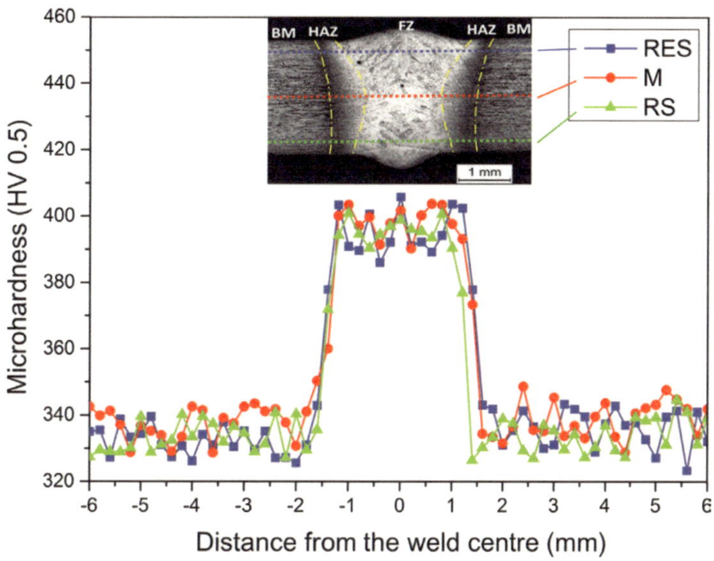

Figure 9. Microhardness profile of the laser beam welded Ti-6Al-4V butt joint measured in the as-welded condition.

3.3.1. Influence of PWHT on microhardness

The variations of average microhardness values for the BM and FZ in the as-welded condition and after PWHT under different conditions are shown in **Figure 10**. The hardness in the FZ was generally higher than that in the BM, but their difference depended strongly on the annealing temperature. The changes in microhardness after conducting PWHT are strictly related to the microstructural changes that occurred during the PWHT. Because the effect of PWHT depends on the initial microstructure, the BM and FZ underwent different transformations during PWHT and will be discussed separately.

Annealing at 540°C for 4 h slightly increased the average microhardness of the BM from 336 ± 8 HV 0.5 in the as-welded condition to 351 ± 10 HV 0.5 after annealing. This unexpected hardening effect upon low-temperature annealing was not evident from OM and SEM observations and can be attributed to precipitation hardening of the α phase by coherent Ti_3Al

particles [1, 28]. During annealing, significant alloy element partitioning takes place; that is, α phase is enriched with α stabilizing elements (Al), and β phase is enriched with β stabilizing elements (V, Fe) owing to diffusion processes. This fact was proved by EDX point analysis. **Figure 11** shows the average content of alloying elements in α and β phases of the BM after PWHT. As shown in **Figure 11**, Al content in the α phase and V and Fe content in the β phase increase with increasing temperature. Coherent α_2 particles can then be precipitated in the alpha phase by ageing owing to increased Al content. This age-hardening effect of the α phase in Ti-6Al-4V by Ti_3Al particles was well documented by Lutjering et al. [28]. In the Ti-6Al-4V alloy, the Ti_3Al solvus temperature is approximately 550°C. Annealing at temperatures lower than 550°C will precipitate α_2 particles, whereas a heat treatment at 600°C will be only a stress-relieving treatment [29]. The latter was in a good agreement with our findings. Heat treatment at 650°C for 1 h led to average microhardness in the BM of 331 ± 10 HV 0.5, which is slightly lower than that in the starting condition. Evidence of this ageing phenomenon is not apparent from optical microscopy images and EBSD results because the size of the α_2 particles is approximately several nanometres as reported in Ref. [28]. The application of transmission electron microscopy (TEM) techniques to study this complex phenomenon should improve identification of the α_2 phase.

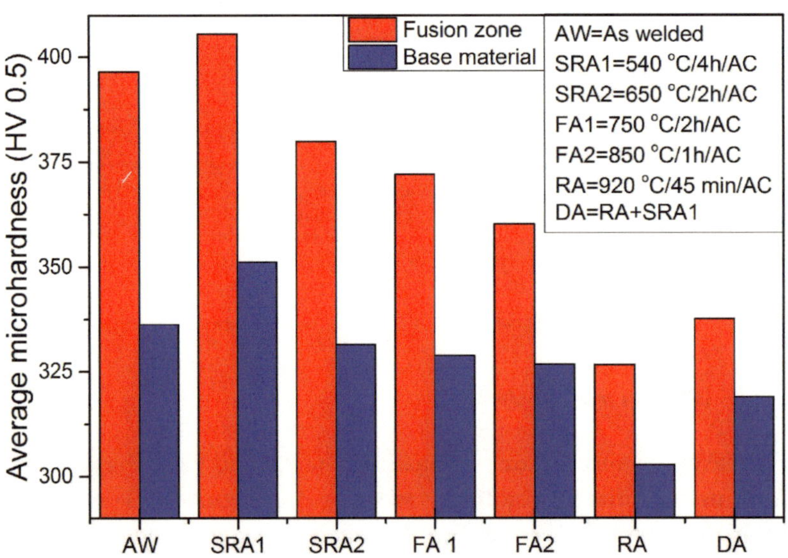

Figure 10. PWHT influence on the average microhardness in the FZ and BM.

As presented in **Figure 10**, annealing in the temperature range of 650–850°C does not affect the average microhardness in the BM significantly, and It gradually decreased with increasing temperature. A substantial decrease in microhardness was observed after recrystallization annealing at 920°C for 45 min. These results are related to grain coarsening in accordance with the Hall-Petch mechanism and are consistent with microstructural observations discussed in the previous section. Because the second step of duplex annealing is equal to SRA1 heat

http://dx.doi.org/10.5772/66178

treatment, this led to hardening of the coarsened structure obtained after RA due to precipitation of α_2 phase as discussed above.

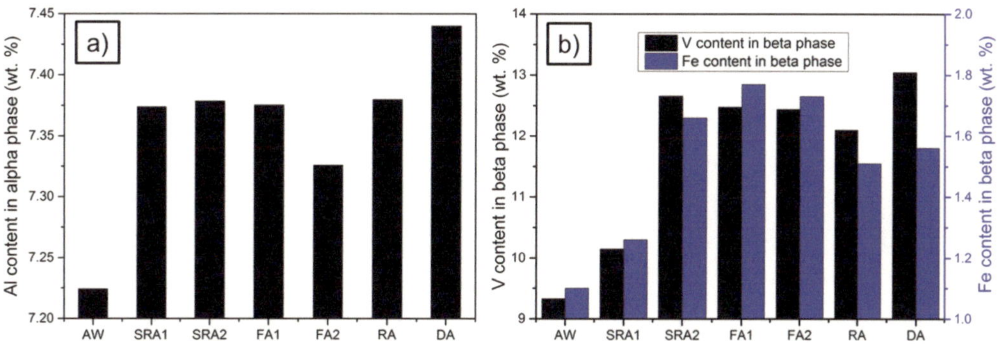

Figure 11. Alloying element partitioning in the BM due to PWHT. (a) Al in α phase and (b) V and Fe in β phase.

The effect of various PWHT on the average microhardness in the FZ has a generally very similar trend as discussed for the BM; however, the mechanisms leading to these results were not the same because the initial microstructures were completely different. The hardening effect observed in the FZ after SRA1 annealing can be attributed to tempering of the martensitic structure. This effect at relatively low temperatures has already been observed by a number of researchers [6, 30, 31]. Because metastable α' martensite is supersaturated in β stabilizers owing to the diffusionless transformation $\beta \rightarrow \alpha'$, upon annealing, it decomposes into $\alpha + \beta$ by precipitation of incoherent β particles at dislocations or β phase layers at plate boundaries [28]. Chesnutt et al. [32] investigated ageing of β-quenched Ti-6Al-4V and used TEM to show microprecipitation of β phase particles in tempered martensite. Precipitation hardening takes place only after SRA1 annealing, whereas heat treatment at 650°C for 2 h leads to partial decomposition of martensite with attendant reduction of microhardness. Starting from the temperature of 750°C, grain coarsening of an acicular microstructure in the FZ is responsible for the gradual reduction of microhardness with increasing temperature (see **Figure 10**). After recrystallization, the annealing microstructure in the FZ was completely transformed into equilibrium coarse lamellar $\alpha + \beta$ morphology with the lowest hardness values nearly equal to that of the BM in the as-received condition. The hardening mechanism that took place after ageing in DA was apparently the same as described above for the BM because after recrystallization, both the BM and FZ consisted of equilibrium $\alpha + \beta$ phases.

3.4. Residual stress analysis

The HDM method gives the average stresses in the area where the material was drilled, that is, in the circle of 0.6 mm in diameter. Because no depth-dependent gradient of residual stresses was observed in the present study, each point is represented by the mean value of residual stresses at the depth of 0.3 mm. The distribution of residual stresses longitudinal and transverse to the weld line across the welding seam is shown in **Figure 12(a)**. High tensile longitudinal stresses up to 650 MPa are produced in the weld itself and the immediately adjacent parent material during solidification. These high tensile residual stresses near the weld are balanced by compressive longitudinal stresses further from the weld line. Transverse residual stresses have a similar profile but are an order of magnitude lower than the longitudinal component with a maximum value not exceeding 50 MPa. Our results are in good agreement with that reported by Cao et al. for laser beam welded Ti-6Al-4V alloy [33].

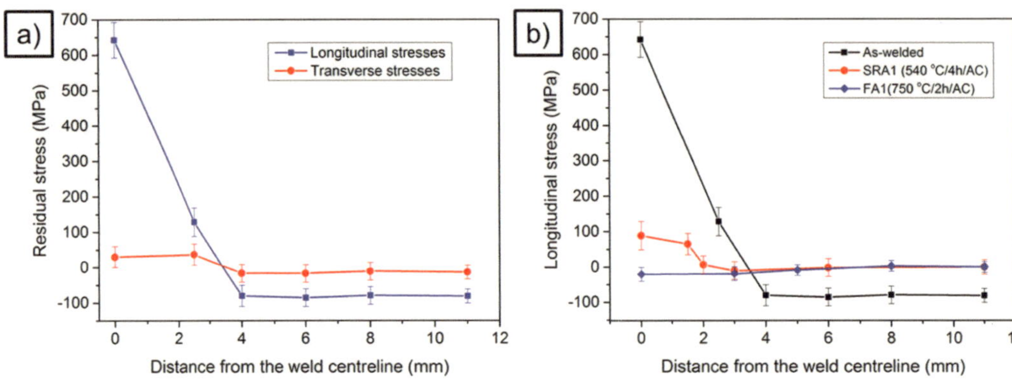

Figure 12. Residual stress distribution in the vicinity of the laser beam welded Ti-6Al-4V butt joint. (a) Residual stresses in the as-welded condition and (b) influence of PWHT on longitudinal residual stress profile.

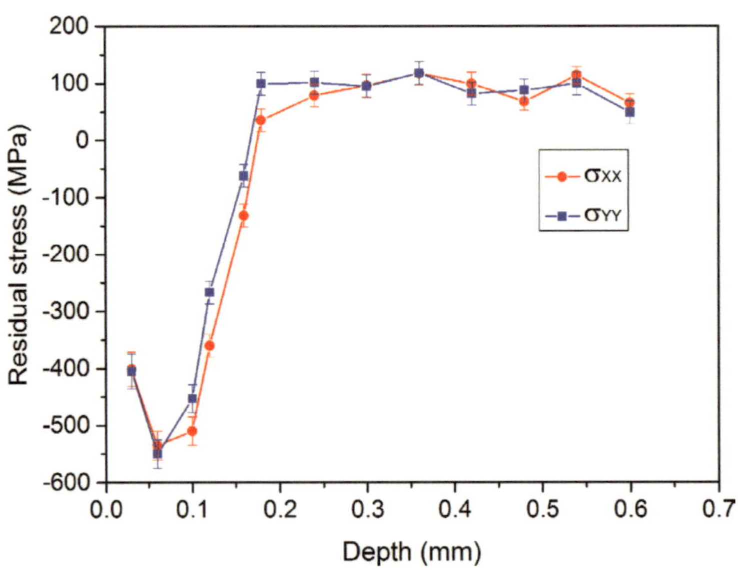

Figure 13. Residual stresses in the BM after milling.

The stress-relieving effect at elevated temperatures due to PWHT is presented in **Figure 12(b)**. Because longitudinal stresses significantly prevail over transverse components, for simplicity, **Figure 12(b)** presents only the evolution of longitudinal stresses upon heat treatment. Annealing at 540°C results in the stress relief from 650 MPa in the centre of FZ to approximately 90 MPa (nearly 85% effect). After heat treatment at 750°C, almost full stress relief was achieved, and PWHT's at higher temperatures completely remove welding-induced residual stresses in the welding seam. It should be kept in mind that the presented results correspond to different specimen geometry from that used in fatigue testing. The specimen size should be considered as an influential factor when evaluating the effect of residual stresses on the fatigue of welded joints [34, 35]. After extracting S–N specimens from the welded plate due to the relatively short width of the gauge length (8 mm), residual stresses are almost fully removed. The latter was confirmed by the HDM technique in the extracted fatigue specimens and does not allow us to make any conclusions about the influence of residual stresses on the fatigue properties of laser beam welded Ti-6Al-4V butt joints. However, in the situations in which the width of the tested specimen is sufficient to keep the residual stresses after cutting or the welding line is parallel to the external stresses, the results presented in this work would be quite useful. Moreover, the stress-relieving effect investigated here is necessary for comprehensive analysis of the PWHT of laser beam welded butt joints. Large-scale specimens must be tested to separately investigate the influence of residual stresses on the fatigue of LBW joints.

Incremental HDM allowed us to investigate the residual stresses that arise in the surface layer after machining. These stresses are of great interest because they significantly affect the unnotched fatigue properties of the material and will be discussed in the last section. **Figure 13** shows a residual stress distribution in the BM after milling. As we can see, high compressive residual stresses up to 550 MPa are formed in the 0.2-mm-thick surface layer. These compressive stresses in the surface layer are balanced by tensile stresses in the bulk material.

3.5. Fatigue testing

The results of room-temperature high cycle fatigue tests are shown in **Figures 14** and **16**. For reference, base material data are also provided in **Figure 14**. Arrows indicate non-failures after 10^7 cycles (run-outs). The curves shown in these plots represent the mean lines corresponding to 50% probability of survival. The results highlight the inherent scatter in fatigue test experiments for titanium alloys [36]. As shown in **Figure 14**, the fatigue limit of the BM in the starting as-received condition is approximately 650 MPa or nearly 65% of the yield strength. Machining both reduces the surface roughness from 3 to 1.5 Rz and results in the formation of compressive residual stresses in the 0.2-mm-thick surface layer (see **Figure 13**). Improved surface quality and introduction of favourable compressive stresses in the near surface region have a beneficial effect on the HCF resistance of the Ti-6Al-4V BM. As seen in **Figure 14**, the fatigue limit increased to 720 MPa after milling the surface of the specimens. This result should be kept in mind when comparing the effect of milling the weld defects on the HCF properties. The S-N curve of the milled BM must be considered as the reference for machined weldments.

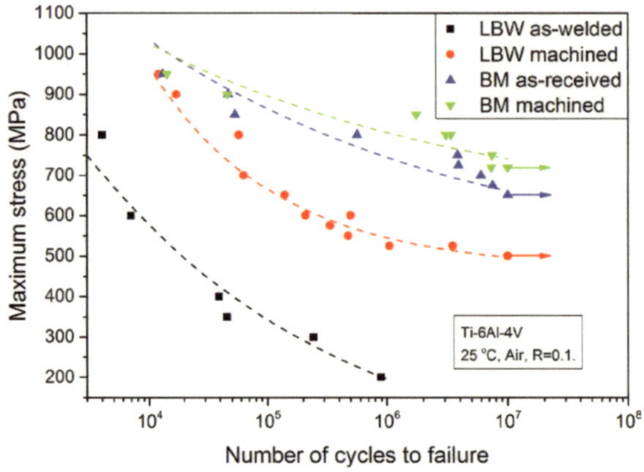

Figure 14. Influence of machining on the fatigue behaviour of the laser beam welded Ti-6Al-4V butt joints.

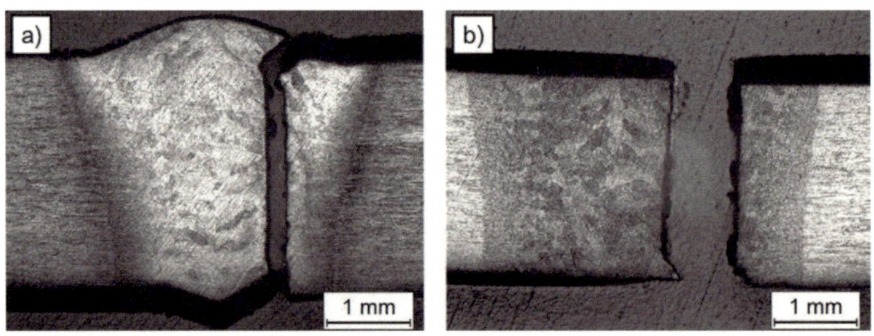

Figure 15. Transverse cross sections of fractured S-N specimens. (a) As-welded condition, 200 MPa, 889,500 cycles and (b) annealed (FA2) and machined, 575 MPa, 4,953,100 cycles.

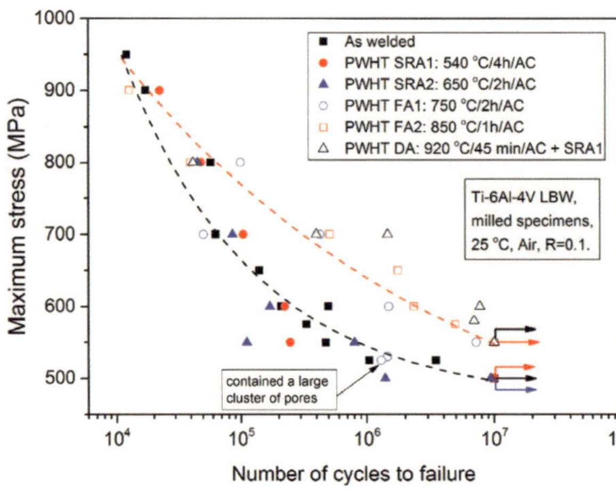

Figure 16. Influence of PWHT on the fatigue behaviour of the laser beam welded Ti-6Al-4V butt joints.

http://dx.doi.org/10.5772/66178

3.5.1. Influence of defects

The S-N curves for the as-welded and machined flush laser beam welded Ti-6Al-4V butt joints are given in **Figure 14**. The fatigue limit after milling the weldment flush with the sheet surface was approximately 500 MPa. This value corresponds to 70% of the base material fatigue limit (also machined). It can be seen (**Figure 14**) that the presence of reinforcements and small underfills significantly deteriorates the fatigue of the laser beam welded butt joints. Geometry imperfections such as underfills and reinforcements play the role of stress concentrators (notches). The failure always occurred in the welding seam initiated at the face or root underfill. **Figure 15** shows typical transverse cross sections of fractured S-N specimens in the as-welded and machined conditions. In the as-welded specimen, the crack started from the weld root and propagated through the FZ perpendicular to the direction of applied stress as shown in **Figure 15(a)**.

Thus, machining the weld reinforcements and underfills flush with the sheet surface can be considered as an easy method to improve the fatigue performance of the laser beam welded butt joints. These results are consistent with the work of Squillace et al. [2]. They showed that the fatigue strength of autogenous laser beam welded Ti-6Al-4V butt joints is strongly influenced by the value of the underfill radius, and the S-N curves shift towards the region of HCF as the value of the underfill radius increases. Improved fatigue strength by partially or totally eliminating the underfills, predicted in the above-mentioned work, was confirmed in the present study. The use of filler wire partly prevented the formation of underfills; however, as seen in **Figure 14**, the synergetic effect of the weld reinforcements and underfills considerably affected the fatigue performance, although the acceptance criteria in terms of geometrical defects were passed.

In the low cycle fatigue (LCF) region, the S-N curve of the milled condition approaches the static strength of the laser beam welded joints, which usually equals the strength of the parent material [2, 5]. The specimen tested at the 950 MPa level of maximum stress was fractured in the base metal. This implies that in the LCF region, the laser beam welded Ti-6Al-4V flush milled butt joints exhibit a BM level of fatigue strength. All other laser beam welded specimens tested in the current work were fractured in the FZ. The typical location of failure in the machined laser beam welded joint is shown in **Figure 15(b)** for the specimen, which endured nearly five million cycles at the 575 MPa level of maximum stress.

In the HCF regime, the S-N curve for the flush milled condition is located lower than that of the base material. The fatigue limit decreased by nearly 31%. This result implies the existence of internal microstructural features or defects deteriorating the fatigue strength of the joint. In experiments with butt welds in the as-welded condition, the stress concentration at the weld toes or roots is much more severe than that due to minor defects existing in the welding zone, and these defects are therefore less important. Thus, geometry features can overshadow the microstructural effects and internal defects. The latter are of primary interest in this work. By removing the stress concentrators from the surface of the welding seam, internal defects become the most important notches in the joint and exhibit their full deleterious effect.

3.5.2. *Influence of PWHT on fatigue*

The influence of PWHT on the fatigue performance of the laser beam welded Ti-6Al-4V butt joints is shown in **Figure 16**. It should be kept in mind that all data given in **Figure 16** represent the flush milled condition. If the annealing temperature is lower than 750°C, the results of the tests lie in the scattering range of the as-welded S-N curve. No significant influence of low-temperature annealing ($T < 750$°C) on fatigue performance was found. Conditions SRA1 and SRA2 had approximately the same fatigue limit of approximately 500 MPa, considering the inherent scatter of fatigue experiments.

Annealing at temperatures above 750°C leads to the slight improvement of fatigue properties. The specimens annealed at 850°C (FA2) and subjected to DA showed the highest value of fatigue limit in the present work: 550 MPa. For comparison, as-welded specimens heat treated at lower temperatures endured less than 500,000 cycles at that level of stress. The fatigue limit for the FA1 condition was not achieved in the current study. The last two FA1 specimens exhibited very low fatigue life compared with the position of other points for this condition. The examination of fracture surfaces revealed the presence of relatively large clusters of pores with the size of approximately 300 μm. These specimens were probably extracted from the region of the plate with worse quality of the welding (run-in or run-outs). Even considering the scatter of the results, the general trend, that annealing at high temperatures (>750°C) increases the fatigue strength of the joint and shifts the S-N curve towards higher values of stresses, can be clearly seen in **Figure 16**.

3.6. Microfractography

Fracture surfaces were studied using OM and SEM to identify the locations of crack initiation, region of stable crack propagation and overload region morphology. After careful fracture surface examination of the broken S-N specimens with an optical microscope, it was concluded that almost 100% of these failures started from internal welding defects, that is, pores with average diameters of approximately 10–100 μm. These pores play the role of structural discontinuities and stress concentrators (notches). The existence of such high stresses in the specimen leads to the initiation of the microcracks in very early periods of fatigue life [37]. **Figure 17** shows the fracture surface overall views of the heat-untreated machined specimen (a) and the specimen after PWHT at 850°C for 1 h (b). Crack initiation sites are clearly seen. The typical distance from the surface to the crack nucleation point is approximately 300–700 μm; however, less frequently, pores appear closer to the surface. Cracks initiated from a single pore are very rare and occur only if the size of the pore is approximately 100 μm. More frequently, the cracks start from the subsurface clusters of pores (usually two or three pores), as shown in **Figure 17(a)**. The distance between cracks in subsurface clusters is typically less than the pore diameter and does not exceed 10 μm. According to AWS D17.1 [19], two or more discontinuities in the welding zone should be treated as one when the spacing between them is less than the dimension of the larger discontinuity. This allows us to consider subsurface clusters of pores as single defects with an approximate size of 150–250 μm.

The area around the crack initiation site is slightly brighter than the region of stable crack growth. This white circle resembles a "fish eye" fracture, common for steels in the ultra-long-life regime, when fatigue fracture origins are mostly at non-metallic inclusions in the subsurface [37]. The "fish eye" in the present study always had a radius equal to the distance from the surface to the crack nucleation site. The difference in the colours is probably caused by cyclic contact of the fracture surfaces in the absence of atmospheric effects within the "fish eye" and by cyclic contact of the fracture surfaces in the presence of atmospheric gases outside the "fish eye" [37].

Figure 17. OM images of fracture surfaces. (a) As-welded condition, 525 MPa, 3,488,300 cycles and (b) PWHT (FA2), 650 MPa, 1,763,300 cycles.

The SEM images with higher magnification of the specimen shown in **Figure 17(b)** revealing the topography of the fracture surface in different zones of crack growth are shown in **Figure 18**. The zone adjacent to the pore (**Figure 18(a)**) is characterized by low values of stress intensity factor and shows fibrous morphology. A comparison of the fracture face with the microstructure (see **Figure 8**) suggests that the elongated fracture features correspond to individual α laths. The crack propagated radially from the pore and was dominated by a transgranular mode of cracking. At high stress intensity factors (**Figure 18(b)**), the fracture topography was mainly characterized by typical fatigue striations and secondary cracks. The overload region exhibited small, shallow dimples, which are indicative of ductile fracture due to microvoid coalescence (**Figure 18(c)**).

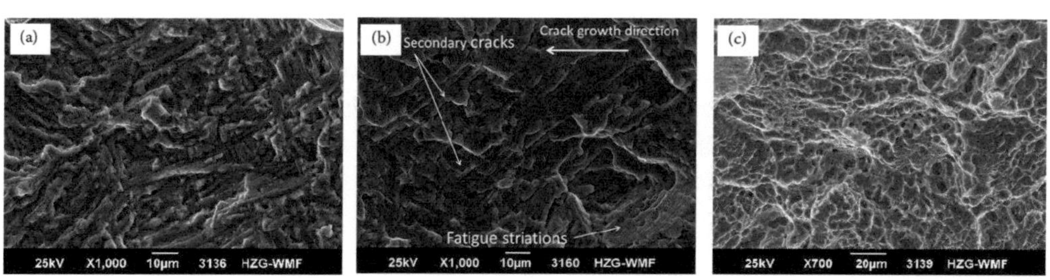

Figure 18. SEM images of fatigue fracture surfaces. PWHT (FA2), 650 MPa, 1,763,300 cycles. (a) Region close to the pore, (b) region of stable crack growth, and (c) overload region, final fracture.

4. Discussion

Because the specimens that were subjected to PWHT were milled thereafter, we can assume that equality of surface roughness was provided, and we can exclude this factor from our further consideration. This allows us to consider microstructural changes as the main factor affecting the fatigue properties of the laser beam welded joints in the present work. The aim of this section is to link the results of mechanical testing with microstructural observations, microhardness measurements and fracture surface topography analysis. As described above, fatigue failure in the laser beam welded joints was always found in the FZ. Thus, the microstructure in this region of the weldment is a crucial factor affecting the fatigue performance of the joints.

The HCF failure process comprises four stages: microcrack initiation, microcrack propagation, macrocrack propagation and final fracture [38]. The border between micro- and macrocrack propagation can be defined when the size of the crack is approximately one order of magnitude larger than the effective microstructural size [39], which is the average grain size (3–6 μm) in the case of an equiaxed microstructure and is the α colony size (10–40 μm, depending on the cooling rate) in the case of a lamellar structure. A number of researchers have shown that crack initiation and microcrack propagation take up to 95% of the high cycle fatigue life [28, 37, 40]. Macrocrack propagation is very fast relative to the first stages and does not play any significant role in unnotched HCF. In the presence of defects, the size and sharpness of notches are of great importance because they determine the size of the plastic zone at the tip of the notch. As already mentioned above, we can consider subsurface clusters of pores as single defects with approximate sizes of 150–250 μm. A sharp notch with a small crack at its tip may be regarded as a crack [37]. Because the transition short/long crack is on the order of 200 μm (10 times the microstructural size) for lamellar microstructures, a crack emanating from the pore in the FZ can already be assumed to be a long one after the initiation period. This brings us to an important conclusion that the HCF of the LBW joints primarily depends on crack initiation from the pore and near-threshold macrocrack propagation.

A comprehensive study on the influence of microstructural variables on near-threshold fatigue behaviour of macrocracks in titanium Ti-6Al-4V was conducted by Yoder et al. [41–43]. Bilinear crack growth rate behaviour was observed, with two distinct branches that independently obey the power law and join together in the transition point (ΔK_T). In the region $\Delta K_{th} < \Delta K < \Delta K_T$ (ΔK_{th} stands for the threshold stress intensity factor), the cyclic plastic zone is less than the effective grain size, and a microstructurally sensitive mode of crack growth occurs that involves crystallographic bifurcation in grains adjacent to the crack plane. In contrast, in the region $\Delta K > \Delta K_T$, the grains within a larger plastic zone deform as a continuum, which results in a microstructurally insensitive, non-bifurcated mode of crack growth. The observed values of ΔK_T were in remarkable agreement with predictions according to the equation $\Delta K_T = 5.5\, \sigma_y$ vd, where σ_y is the yield strength of material, and d is the effective microstructural size. In the case of titanium alloys, the Hall–Petch relation is relatively weak [44], so the d term in the above-mentioned equation dominates, leading to the increase of ΔK_T with increasing grain size. The inverse dependence of fatigue crack growth rates upon grain size was directly related to the microstructurally sensitive mode of crack growth because larger bifurcated cracks occur with increasing grain size.

The most influential microstructural parameter on the mechanical properties of lamellar (platelet) microstructures is the α colony size (packet size) because it determines the effective slip length in lamellar structures [28, 41]. This parameter should be considered as the effective microstructural size for lamellar morphology in the FZ. With increasing α colony size, the unnotched fatigue strength and yield stress decrease because smooth HCF strength depends primarily on the resistance to crack nucleation and microcrack propagation. However, if macrocracks or sharp notches already exist in the material, a coarse lamellar microstructure is more beneficial for fatigue performance because increased effective slip length retards fatigue crack propagation due to increased crack front roughness. The trade-off between decreased strength and increased fracture toughness makes the coarse lamellar microstructure less sensitive to notches and more advantageous for usage in applications, in which notched fatigue performance is the crucial factor.

Annealing of the laser beam welded Ti-6Al-4V butt joints at temperatures up to 650°C is insufficient for full martensite decomposition into an equilibrium lamellar $\alpha + \beta$ structure in the FZ and recrystallization in the BM. The width of individual α lamellae and the average α colony size remained almost the same after heat treatments at low temperatures. Hence, mechanical properties also should have remained approximately the same. We can conclude that the low-temperature annealing does not affect the HCF performance of the laser beam welded Ti-6Al-4V butt joints.

As described above, starting from the temperature of 750°C, the metastable martensitic structure in the FZ transforms to equilibrium platelet $\alpha + \beta$ morphology, as shown in **Figure 8**. With increasing temperature during PWHT, the width of individual lamellae and α colony size increases with a commensurate increase in the effective slip length and a corresponding decrease in the yield stress. A coarse lamellar structure with lower density of defects after annealing at high temperatures has lower strength and higher ductility than martensitic morphology [6, 45, 46]. This was indirectly confirmed by the decreased microhardness in the FZ after PWHT at high temperatures. Although the static strength of the joint decreased slightly, more ductile and softer material in the FZ was more beneficial for the HCF of laser beam welded joints than a hard martensitic structure. High-temperature annealing reduced the notch sensitivity of the FZ, and internal defects were less detrimental for fatigue properties than in a martensitic structure. Increased α colony size and consequently larger effective slip length leads to a more bifurcated crack growth profile and increased crack propagation resistance of the material. This result is consistent with the works of Yoder et al. [41–43] and Lütjering and Williams [28].

To verify our assumptions, transverse cross sections of fractured specimens were made. Grinding and polishing was performed to the plane containing the pore to compare the crack front roughness in different conditions. As shown in **Figure 19**, the coarser lamellar microstructure displayed a more tortuous and deflected crack path than the finer-scale martensitic microstructure in the region adjacent to the pore, where the crack growth is sensitive to the microstructure. The effective slip length in the case of annealed material is the α colony size, but in case of martensitic morphology, it is the width of individual α plates. Increased crack path tortuosity leads to enhanced crack deflection and a resulting increase in the FCP resistance and the overall fatigue performance.

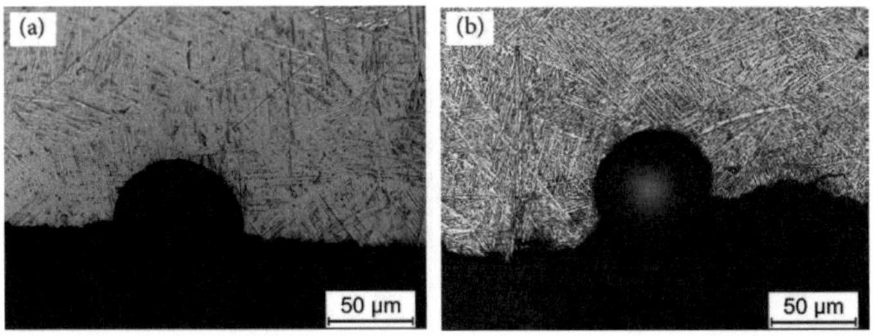

Figure 19. Comparison of crack front roughness profiles in the zone adjacent to the pore. (a) As welded condition and (b) annealed at 850°C for 1 h.

The effect of grain size on the notch sensitivity is well known for steels. Murakami [37] investigated microcrack propagation starting from artificial holes with diameters of 35–500 μm in steels. They concluded that defects smaller than a critical size are non-damaging (not detrimental) to fatigue strength, and the critical size is smaller for materials having higher static strength. Harder material is more sensitive to notches and defects. A larger decrease in fatigue strength for materials of higher static strength was found. This trend is generally in reasonable agreement with our findings, although we investigated different materials. Based on the results for steels, we can also assume the existence of non-damaging permissible defects for titanium alloys because micromechanisms and models for the behaviour of cracks emanating from sharp notches do not depend on the material. However, further investigations must be carried out to prove this assumption. If the dependence of critical size on the microstructure of the FZ is obtained, an obvious way to exclude the detrimental effect of pores on fatigue is to provide a welding technique leading to smaller pores than the critical size. However, it should be kept in mind that in the absence of pores or if the size of defects is lower than the critical size, PWHT will have an opposite effect on fatigue; that is, fatigue strength will be reduced after high-temperature annealing. This was shown in the work of Babu et al. [7] for electron beam welded joints. The fatigue cracks in their work originated at the surface, implying that internal defects were insignificant. PWHT at temperatures of approximately 900°C reduced the fatigue strength of the joints at $2 \cdot \times 10^6$ cycles compared to that of the as-welded condition and those annealed at lower temperatures.

5. Conclusions

Fully penetrated Ti-6Al-4V butt joints were produced by LBW with filler wire. The effect of PWHT was analysed in terms of microstructural features, microhardness, residual stress distribution and fatigue performance. The following conclusions can be drawn:

http://dx.doi.org/10.5772/66178

1. The microstructure of the laser beam welded seam in the as-welded condition was characterized by strong inhomogeneity due to high cooling rates imposed by the welding process. FZ consists of a metastable martensitic structure formed by the diffusionless $\beta \rightarrow \alpha'$ transformation. High spatial gradients of the microstructure and microtexture in the HAZ result in mechanical properties mismatch between the BM and FZ, which was indirectly confirmed by the microhardness profile.

2. Significant grain coarsening was found after PWHT at temperatures higher than 750°C owing to recrystallization processes in the BM and martensite decomposition in the FZ. After high-temperature annealing, the BM microstructure has much fewer lamellar regions, and in the FZ, α at prior β grain boundaries appears, and pronounced α colonies are formed.

3. In the as-welded condition, microhardness values in the FZ were 18% higher than those in the BM. Annealing at 540°C led to the hardening of both the FZ owing to the tempering of martensite and the BM owing to precipitation of α_2 phase. Annealing in the temperature range of 650–920°C promotes grain coarsening in the BM and martensite decomposition in the FZ. As a result, microhardness decreased with increasing temperature. Duplex annealing allowed the microhardness mismatch between the FZ and BM to be minimized to 6%.

4. Annealing at 540°C for 4 h had an effect of relieving approximately 85% of stress. PWHT at higher temperatures completely eliminated welding residual stresses. Machining of the weld defects produces compressive residual stresses in the 0.2-mm-thick surface layer, which are beneficial for fatigue.

5. The fatigue limit of a laser beam welded machined butt joint was 500 MPa or 70% of the milled BM fatigue limit. Stress concentration due to underfills and reinforcements significantly deteriorates the fatigue performance. PWHT at temperatures higher than 750°C increase the fatigue limit by approximately 10% up to 550 MPa. The failures of machined samples started from the subsurface clusters of pores. Increased crack front roughness in the microstructure-sensitive region of crack propagation and increased critical size of non-damaging defects are supposed to be the most influential factors leading to improved fatigue properties of laser beam welded Ti-6Al-4V butt joints subjected to PWHT. A slight decrease in static strength is the main offset for this method.

Author details

Fedor Fomin*, Volker Ventzke, Falk Dorn, Nikita Levichev and Nikolai Kashaev

*Address all correspondence to: fedor.fomin@hzg.de

Institute of Materials Research, Materials Mechanics, Helmholtz-Zentrum Geesthacht, Geesthacht, Germany

References

[1] Boyer R, Welsch G, Collings EW, editors. Materials Properties Handbook: Titanium Alloys. 1st ed. Materials Park, OH: ASM International; 1994. pp. 483–633.

[2] Squillace A, Prisco U, Ciliberto S, Astarita A. Effect of welding parameters on morphology and mechanical properties of Ti-6Al-4V laser beam welded butt joints. Journal of Materials Processing Technology. 2012;212(2):427–436. DOI: 10.1016/j.jmatprotec.2011.10.005

[3] Kabir ASH, Cao X, Medraj M, Wanjara P, Cuddy J, Birur A. Effect of welding speed and defocusing distance on the quality of laser welded Ti-6Al-4V. In: Materials Science and Technology; October 17–21, 2010; Houston, Texas; 2010. pp. 2787–2797.

[4] Ahn J, Chen L, Davies CM, Dear JP. Parametric optimisation and microstructural analysis on high power Yb-fibre laser welding of Ti–6Al–4V. Optics and Lasers in Engineering. 2016;86:156–171. DOI: 10.1016/j.optlaseng.2016.06.002

[5] Kashaev N, Ventzke V, Fomichev V, Fomin F, Riekehr S. Effect of Nd:YAG laser beam welding on weld morphology and mechanical properties of Ti–6Al–4V butt joints and T-joints. Optics and Lasers in Engineering. 2016;86:172–180. DOI: 10.1016/j.optlaseng.2016.06.004

[6] Kabir ASH, Cao X, Gholipour J, Wanjara P, Cuddy J, Birur A, Medraj M. Effect of postweld heat treatment on microstructure, hardness, and tensile properties of laser-welded Ti-6Al-4V. Metallurgical and Materials Transactions A. 2012;43(11):4171–4184. DOI: 10.1007/s11661-012-1230-5

[7] Babu NK, Raman SGS, Srinivasa Murthy CV, Reddy GM. Effect of beam oscillation on fatigue life of Ti-6Al-4V electron beam weldments. Materials Science and Engineering A. 2007;471:113–119. DOI: 10.1016/j.msea.2007.03.040

[8] Tsai CJ, Wang LM. Improved mechanical properties of Ti–6Al–4V alloy by electron beam welding process plus annealing treatments and its microstructural evolution. Materials and Design. 2014;60:587–598. DOI: 10.1016/j.matdes.2014.04.037

[9] Edwards P, Ramulu M. Fatigue performance of friction stir welded Ti–6Al–4V subjected to various postweld heat treatment temperatures. International Journal of Fatigue. 2015;75:19–27. DOI: 10.1016/j.ijfatigue.2015.01.012

[10] ASTM E384-11, Standard Test Method for Knoop and Vickers Hardness Of Materials, ASTM International, West Conshohocken, PA, 2011. DOI: 10.1520/E0384-11E01.

[11] ASTM E837-13a, Standard Test Method for Determining Residual Stresses by the Hole-Drilling Strain-Gage Method, ASTM International, West Conshohocken, PA, 2013. DOI: 10.1520/E0837.

[12] Steinzig M, Ponslet E. Residual stress measurement using the hole drilling method and laser speckle interferometry, Part I. Experimental Techniques. 2003;27(3):43–46.

[13] Schajer G, Steinzig M. Full-field calculation of hole-drilling residual stresses from ESPI data. Experimental Mechanics. 2005;45(6):526–532. DOI: 10.1007/BF02427906

[14] Schajer GS, Prime MB. Use of inverse solutions for residual stress measurements. Journal of Engineering Materials and Technology. 2006;128(3):375–382.

[15] ASTM E 466-15, Standard Practice for Conducting Force Controlled Constant Amplitude Axial Fatigue Tests of Metallic Materials, ASTM International, West Conshohocken, PA, 2015. DOI: 10.1520/E0466-15

[16] Blackburn JE, Allen CM, Hilton PA, Li L, Hoque MI, Khan AH. Modulated Nd:YAG laser welding of Ti–6Al–4V. Science and Technology of Welding and Joining. 2010;15(5): 433–439. DOI: 10.1179/136217110X12731414739718

[17] Blackburn J, Allen CM, Khan A, Hilton PA, Li L. Dual focus Nd:YAG laser welding of titanium alloys: effect on porosity formation. Lasers in Engineering. 2012;22(5–6):319–336.

[18] Cao X, Kabir ASH, Wanjara P, Gholipour J, Birur A, Cuddy J, Medraj M. Global and local mechanical properties of autogenously laser welded Ti-6Al-4V. Metallurgical and Materials Transactions A. 2014;45:1258–1272. DOI: 10.1007/s11661-013-2106-z

[19] AWS: Specification for Fusion Welding for Aerospace Application, D17.1, Miami, OH: American Welding Society; 2001.

[20] DIN EN 4678:2012-01. Aerospace series – Weldments and brazements for aerospace structures – Joints of metallic materials by laser beam welding – Quality of Weldments, European Committee for Standardization, Brussels, 2011.

[21] Chaturvedi MC, editor. Welding and Joining of Aerospace Materials. Woodhead Publishing Ltd; 2012. pp. 94–102

[22] Blackburn J, Allen C, Hilton P, Li L. Nd:YAG laser welding of titanium alloys using a directed gas jet. Journal of Laser Applications. 2010;22(2):71–78.

[23] Ayman A. Salem. Texture separation for a/b titanium alloys. In: Schwartz A.J, Kumar M, Adams B.L, Field D.P, editors. Electron Backscatter Diffraction in Materials Science. 2nd ed. Springer; 2009; pp. 317–327. DOI: 10.1007/978-0-378-88136-2

[24] Peters M, Williams JC. Microstructure and Mechanical properties of a welded (α+β) Ti Alloy. Metallurgical Transactions A. 1984;15:1589–1596.

[25] Ahmed T, Rack HJ. Phase transformations during cooling in α+β titanium alloys. Materials Science and Engineering A. 1998;243(1–2):206–211. DOI: 10.1016/S0921-5093(97)00802-2

[26] Sallica-Leva E, Caram R, Jardini AL, Fogagnolo JB. Ductility improvement due to martensite a' decomposition in porous Ti-6Al-4V parts produced by selective laser

melting for orthopedic implants. Journal of the Mechanical Behavior of Biomedical Materials. 2016;54:149–158. DOI: 10.1016/j.jmbbm.2015.09.020

[27] Kashaev N, Ventzke V, Horstmann M, Riekehr S, Yashin G, Stutz L, Beck W. Microstructure and mechanical properties of laser beam welded joints between fine-grained and standard Ti-6Al-4V sheets subjected to superplastic forming. Advanced Engineering Materials. 2015;17(3):374–382. DOI: 10.1002/adem.201400202

[28] Lütjering G, Williams JC. Titanium. 1st ed. Berlin, Heidelberg: Springer-Verlag; 2003.

[29] Lutjering G. Influence of processing on microstructure and mechanical properties of (α+β) titanium alloys. Materials Science and Engineering A. 1998;243(1–2):32-45.

[30] Thomas G, Ramachandra V, Ganeshan R, Vasudevan R. Effect of pre- and post-weld heat treatments on the mechanical properties of electron beam welded Ti-6Al-4V alloy. Journal of Materials Science. 1993;28:4892–4899.

[31] Wang SQ, Liu JH, Chen DL. Effect of post-weld heat treatment on the fatigue properties of dissimilar titanium alloy joints. In: 13th International Conference on Fracture; June 16–21, 2013; Beijing, China. DOI: 10.4028/www.scientific.net/AMR.891–892.1539

[32] Chesnutt JC, Rhodes CG, Williams JC. Relationship between mechanical properties, microstructure, and fracture topography in α+β titanium alloys. In: 78th Annual Meeting of the ASTM. The Symposium on Fractography – Microscopic Cracking Processes, ASTM STP 600, American Society for Testing and Materials. 1976, 99–138.

[33] Cao Z, Che Z, Zou S. The residual stress distribution and fatigue property of TC4 laser-welded joint treated by laser shock peening. In: 5th International Conference on Information Engineering for Mechanics and Materials; January 2015; Atlantis Press; pp. 631–635. DOI: 10.2991/icimm-15.2015.119

[34] Masubuchi K. Analysis of Welded Structures. Oxford/New York: Pergamon Press; 1980.

[35] Gurney TR. Fatigue of Welded Structures. 2nd ed. Cambridge: Cambridge University Press; 1979. 456 p.

[36] Schijve J. Fatigue of Structures and Materials. 2nd ed. Springer; 2009. pp. 373–394.

[37] Murakami Y. Metal Fatigue: Effects of Small Defects and Nonmetallic Inclusions. 1st ed. Oxford: Elsevier; 2002. 369 p.

[38] Anderson TL. Fracture Mechanics: Fundamentals and Applications. 2nd ed. CRC Press LLC; 1995. 688 p.

[39] Suresh S, Ritchie RO. Propagation of short fatigue cracks. International Metals Reviews. 1984;29(6):445–475.

[40] Eylon D, Pierce CM. Effect of microstructure on notch fatigue properties of Ti-6Al-4V. Metallurgical Transactions A. 1976;7:111–121.

[41] Yoder GR, Cooley LA, Crooker TW. Fatigue crack propagation resistance of beta-annealed Ti-6Al-4V alloys of differing interstitial oxygen contents. Metallurgical Transactions A. 1978;9A:1413–1420.

[42] Yoder GR, Cooley LA, Crooker TW. Quantitative analysis of microstructural effects on fatigue crack growth in Widmanstatten Ti-6Al-4V and Ti-8Al-1Mo-1V. Engineering Fracture Mechanics. 1979;11:805–816.

[43] Yoder GR, Cooley LA, Crooker TW. Observations on microstructure sensitive fatigue crack growth in a Widmanstatten Ti-6Al-4V alloy. Metallurgical Transactions A. 1977;8A:1737–1743.

[44] Yoder GR, Cooley LA, Crooker TW. A critical analysis of grain-size and yield-strength dependence of hear-threshold fatigue crack growth in steels. In: Fracture mechanics: 14th Symposium – Volume I; ASTM STP 791; 1983. p. 361.

[45] Fan Y, Shipway PH, Tansley GD, Xu J. The effect of heat treatment on mechanical properties of pulsed Nd:YAG welded thin Ti-6Al-4V. Advanced Materials Research. 2011;189–193:3672–3677.

[46] Cao X, Jahazi M, Marya S, Birur A. Effect of post-weld heat treatment on Nd:YAG laser welding Ti6Al4V alloy quality. Materials Science Forum. 2010;638–642:3655–3660. DOI: 10.4028/www.scientific.net/MSF.638-642.3655

Chapter 7

Grain Boundary Effects on Microstructural Stability of Nanocrystalline Metallic Materials

Xiaofei Zhu, Guangping Zhang and Cheng Yan

Additional information is available at the end of the chapter

http://dx.doi.org/10.5772/66426

Abstract

Grain boundaries play an important role in dictating the mechanical and physical properties of nanocrystalline (NC) materials because of the increased volume fraction of intercrystalline components as the grain size decreases. In general, grain boundaries have a high energy level and there exists a thermodynamic driving force to reduce the overall area of grain boundaries through grain coarsening, making NC material systems intrinsically unstable. Recent investigations also indicate that mechanical deformation can promote grain growth in NC material even at the cryogenic temperatures. In this chapter, first, the current investigation on the grain boundary structures of NC metallic materials is briefly reviewed and then the state-of-the-art of experimental results on the microstructural stability during deformation processes is discussed. Finally, several key issues for improving the microstructure stability of NC metallic materials and possible future work are discussed.

Keywords: nanocrystalline, grain boundary, microstructural stability, deformation mechanism, fracture

1. Introduction

A grain boundary (GB) in the homophase polycrystalline metal is an interface between two crystals of the same crystal structure [1, 2]. According to the dimensional scale of defects in polycrystalline materials, the GB is classified as a planar defect. GBs play an important role in the mechanical properties of polycrystalline metals [3, 4]. For bulk polycrystalline metals, the GB is considered to be stable during the plastic deformation processes and acts as a sink, source or obstacle to the dislocations [1]. With the decreasing microstructural length scale of polycrystalline materials, the volume fraction of atoms residing in or near the grain boundaries increases. Experimental results indicate that GB width (δ) is approximately 0.5 nm for face-centered cubic (fcc) NC alloys and slightly larger than 1.0 nm for body-centered cubic (bcc) NC alloys [3, 5, 6]. Assuming the grains have the shape of spheres, the volume fractions of inter-crystal regions and GB as a function of grain size (d) are shown in **Figure 1**. It can be seen from **Figure 1** that for nanostructured materials with grain size of 5 nm, nearly 50% of atoms will reside in or near the GB [7]. Therefore, nanostructured materials can be considered to compose of two parts: the core crystallites and a network of intercrystal regions (grain boundaries, triple junctions, etc.) [8]. In NC materials, grain boundary mediated processes, such as emission and absorption of dislocations by grain boundaries, grain rotation, and GB sliding will dominate the plastic deformation as the grain size is smaller than a certain critical value [9]. The properties of the nanostructure materials are thus determined not only by their reduced microstructural length scale, but also by the nature of their GB structures [10, 11]. Due to high volume fraction of the GB in NC materials, thermodynamic driving force exists to drive GB migration which results in the low stability of NC materials. The main objective of this chapter is to provide a comprehensive review of the experimental and simulation results on the microstructure instability of NC materials system under various deformation conditions.

This chapter is structured as follows: Section 2 describes the basic methods established in the field of GB structure to describe the GB structure. The section that follows addresses the GB

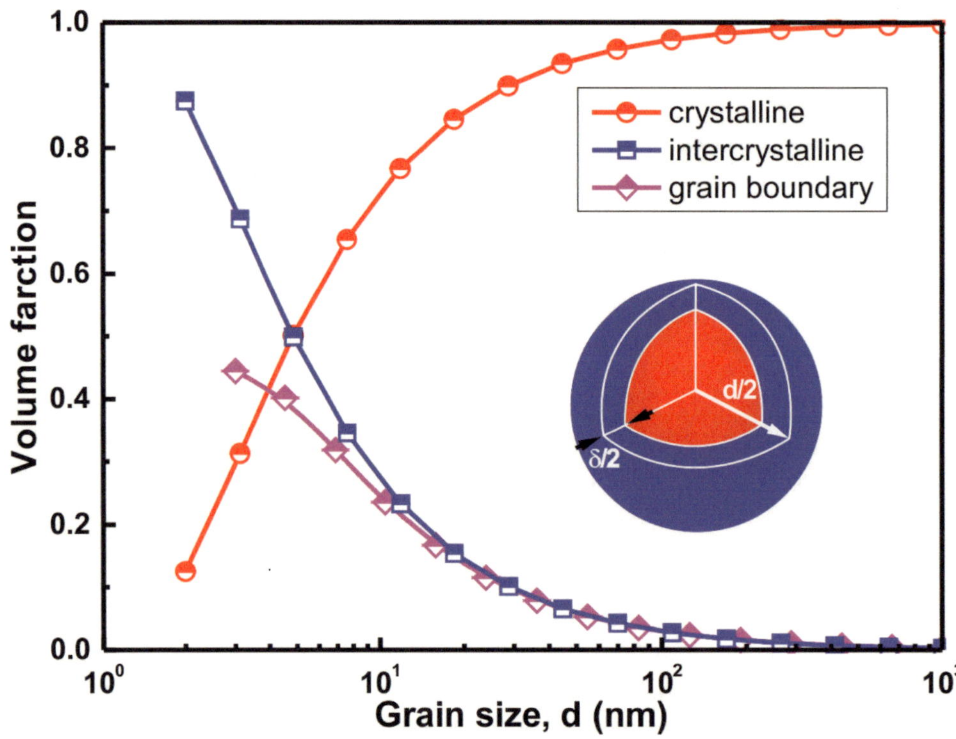

Figure 1. The contribution of different microstructural elements to the volume fraction as a function of grain size (d), assuming a grain-boundary thickness (δ) of 1 nm.

http://dx.doi.org/10.5772/66426

stability phenomena in NC materials, the GB effect on the monotonic and cyclic deformation processes of NC materials. Finally, the application method to enhance the microstructural stability of NC and potential investigations in the future are discussed.

2. Grain boundary structure

For understanding the properties of NC materials, it is a prerequisite to have a detailed knowledge of the GB structure from the atomic (local) scale to the microstructural scale. A considerable effort has been made to understand the GB structure of polycrystalline materials, since the 1950s. To describe the GB crystallographically, a number of parameters must be defined. From a macroscopic perspective, a planar GB between two adjacent grains has five degrees of freedom. Four degrees of freedom are accounted for the crystallographic orientation of the rotation axis and the normal of the GB plane. The fifth is defined as the misorientation angle (θ) [12]. There are several criteria to classify the GB. According to θ value, GBs are typically classified to low-angle boundaries with $\theta \leq 15°$ and high-angle boundaries with $\theta > 15°$ [13]. The coherency of homophase low-angle GBs can be described through the dislocation model [14]. The degree of coherency is related to the spacing of misfit dislocations within the GB. Based on the relative orientation of the rotation axis and the GB plane normal, the GB can be classified as a tilt and twist boundary. If the rotation axis is perpendicular to the GB plane normal, this GB is called a tilt boundary, whereas if the rotation axis is parallel to the GB normal, the GB is defined as a twist boundary. Although tilt and twist GBs occupy only a small fraction of the GB phase space, they are frequently observed experimentally, which suggests that they are energetically favored over other types of GBs [15].

Numerous theoretical efforts have been made to characterize the GB structures. The widely used models to analyze and predict the atomistic structures of GB include the coincidence site lattice (CSL) model and the structural unit (SU) model. In the CSL model, a coincidence index (Σ) is the ratio of the volume of the CSL cell to that of the lattice unit cell. The reciprocal value of the Σ represents the fraction of the lattice points belonging to the abutting crystal. GB with lower value of Σ contains a higher density of coincident sites and is expected to have low energy. While in the SU model, those GBs with specific misorientation angles called favored GB is constituted only from one type of structural units. Any intermediary GB between two favored GBs can be described by a linear combination of SUs comprising one or several neighboring favored boundaries [16, 17]. Twenty-one <1 1 0> symmetric tilt GBs are investigated by Rittner and Seidman with atomistic simulations, using an embedded-atom method potential for low stacking-fault energy fcc metal [18]. They found that the favored boundaries are the $\Sigma = 1$ (0 0 1), $\Sigma = 27$ (1 1 5), $\Sigma = 11$ (1 1 3), $\Sigma = 3$ (1 1 1), $\Sigma = 9$ (2 2 1) and the $\Sigma = 1$ (1 1 0) interfaces. The structural units associated with each of these boundaries are denoted by A–E, as shown in **Figure 2**. To reduce the number of distinct SU, distortions exceeding 15% are occasionally permitted in the SU of the GB region.

The advent of transmission electron microscopy (TEM), especially in the development of high-resolution transmission electron microscopy (HRTEM) has provided us with a very powerful tool to explore the atomic structure of internal interfaces. **Figure 3** shows the HRTEM

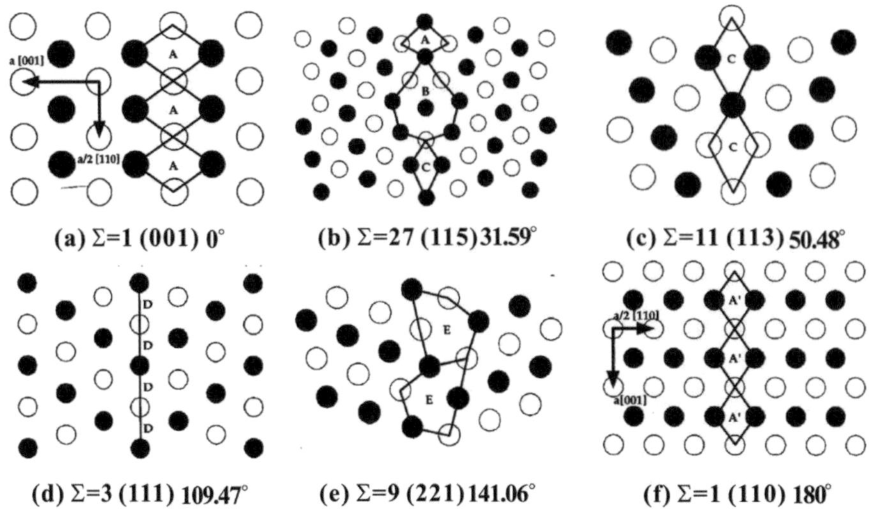

Figure 2. Equilibrium bicrystal interface structures of <1 1 0> symmetric tilt boundaries and two perfect crystal orientation [18]. The structures are viewed along the tilt axis [1 1 0], with the open and filled circles indicating atomic positions in alternate (2 2 0) planes [18].

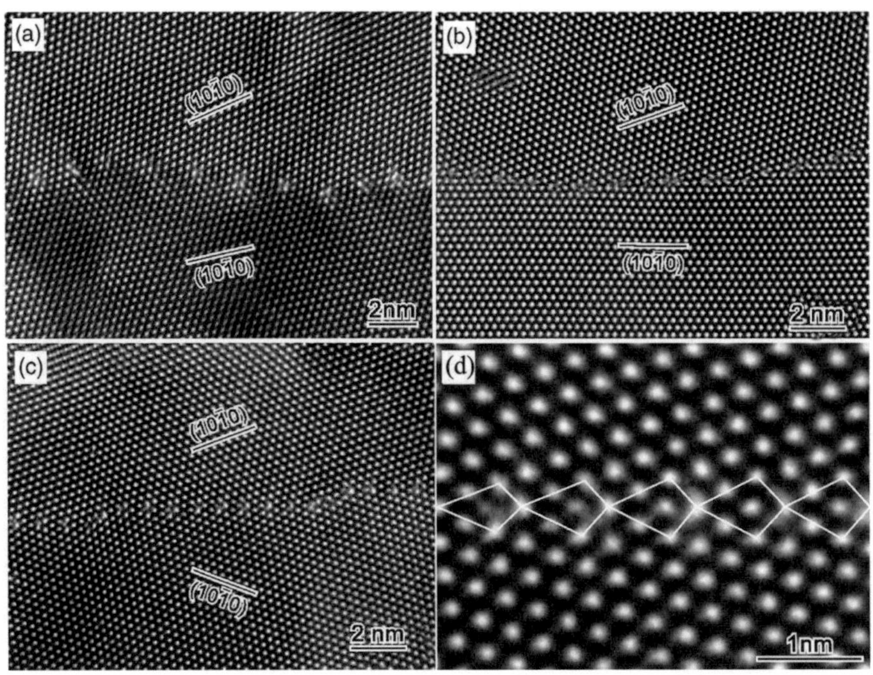

Figure 3. HRTEM images of [0 0 0 1]-tilt grain boundaries in ZnO bicrystals. (a) 10.6 ± 0.1° boundary composed of a dislocation array. (b) 20.1 ± 0.2° near Σ = 7 boundary having a facet structure. (c) 20.0 ± 0.2° near Σ = 7 boundary with a symmetric structure. (d) Higher magnification image of (c) with the boundary core structural units represented by quadrilaterals [19].

micrographs of the atomic structure of ZnO thin films [19]. It is evident that the low-angle GB is composed of edge dislocation along the GB plane as shown in **Figure 3(a)**. **Figure 3(d)** is the higher magnification of **Figure 3(c)**, having a periodic and nearly mirror symmetric character as clarified by drawings of SUs. The HRTEM greatly contributes to the experimental verification of the basic concepts of atomic structure of the GB [12]. TEM characterization results indicate that the GB of NC material is essentially the same as that of the coarse-grained materials. It should be noted that GB structures of NC materials are highly dependent on the alloy composition and processing steps in the manufacturing of NC materials.

3. Microstructural stability

3.1. Grain growth under monotonic deformation

Although NC materials show superior mechanical properties as compared to the coarse-grained counterpart, their application has been severely limited by their microstructural instability under the monotonic mechanical deformation [20–33]. For pure NC metallic materials, the grain coarsening process often occurs even at ambient temperatures [34–36]. Ames et al. reported on the observation of the room temperature grain growth in high-purity NC Pd with an initial grain size of about 10 nm [35]. They found a transition from an initially self-similar slow growth to abnormal grain growth. However, they argued that abnormal grain growth is a transient state since a monomodal grain size distribution was observed in the late stage of coarsening. Discontinuous grain growth in NC materials during the deformation processes seems to be a common phenomenon. Zhang et al. examined the effect of the temperature and sample purity on the grain coarsening behavior and found that many of the grains under the indenter have grown to several hundred nanometers while the unindented microstructure remains unchanged [20, 28], as shown in **Figure 4**. Due to the complex and large stress/strain field of the Vickers indenter is not yet known, the detailed mechanism of the grain growth is not known. However, the increased rate of growth at cryogenic than at the room temperature indicates that the growth is primarily mechanical, not diffusion-driven. In order to elucidate the effect of stress and strain on mechanically induced grain growth, Rupert et al. fabricated specimens with specially designed stress and strain concentrators to reveal the relative importance of these parameters on grain growth [24]. Statistical results of grain size in horizontal-hole specimens showed that grain growth occurred at both high strain region and high stress region. However, grain size was greater in the high stress region, indicating that the grain growth is driven by stress. Statistics results of grain size of angled-hole specimens demonstrated that grain growth was scaled with shear stress. Gianola et al. have investigated the tensile mechanical properties of 180-nm-thick NC Al films with grain size about 40 nm [22]. The specimens that exhibit high strength maintain their NC microstructure, the limited elongation, and the dramatic strain softening. By contrast, specimens that undergo discontinuous grain growth show intermediate strengths and the unexpected development of a region of extended plasticity. Statistical results of grain size outside of the deformed region of specimens that exhibited the grain growth are similar to the initial state indicating that the grain growth is directly tied to the applied stress or deformation in the sample [37].

Figure 4. TEM images of Cu sample after indentation tests (a) in a region away from the indents, and (b) inside an indent made at room temperature with a dwell-time of 30 min [20].

3.2. Grain growth under fatigue test

Grain growth is also observed during the cyclic deformation of NC materials [38–50]. Early fatigue experiments by Witney et al. on the NC Cu reported modest grain growth due to pull-pull cyclic loading [38]. They found that the protrusions stick out on the order of a micrometer which is far greater than the initial grain size, and extend several microns parallel to the surface, similar to extrusions formed during the fatigue test of a coarse-grained Cu. Detailed TEM studies of NC Ni-Fe alloy under cyclic deformation demonstrates that the grain coarsening is accompanied by the fatigue crack growth. Stress concentration at the crack tip causes the lattice reorientation under the cyclic deformation [41]. GB dislocations play a critical role in grain rotation and in the formation of subgrain in larger grains. Boyce and Padilla have reported the fatigue crack initiation and growth behavior in NC Ni, Ni-Mn, and Ni-Fe alloy [42]. They found localized regions of grain growth during fatigue loading. Coarsened Ni grains did not favor any particular orientation, while Ni-0.5Mn coarse grains showed a <1 1 0> preferred orientation. Their observations also suggest that grain stability is an important factor affecting the crack initiation and propagation process in these NC alloys. Meirom et al. reported fatigue-induced grain coarsening during crack propagation in NC Pt films with a strong <1 1 1> texture [43, 44]. They found a clear evidence of increased grain size in the crack wake and ahead of the crack tip. Coarsened grains underwent a nearly one order of magnitude increase in size compared with the as-received one. They also found that many of the grain boundaries in the Pt films are "low-angle" in character and the grain coarsening by the annihilation of low-angle GB through the dislocation slip near the crack tip [43, 44]. Recently, Zhang et al. investigated the fatigue behavior of 100-nm-thick NC Cu film on a polymer substrate [49]. They found that

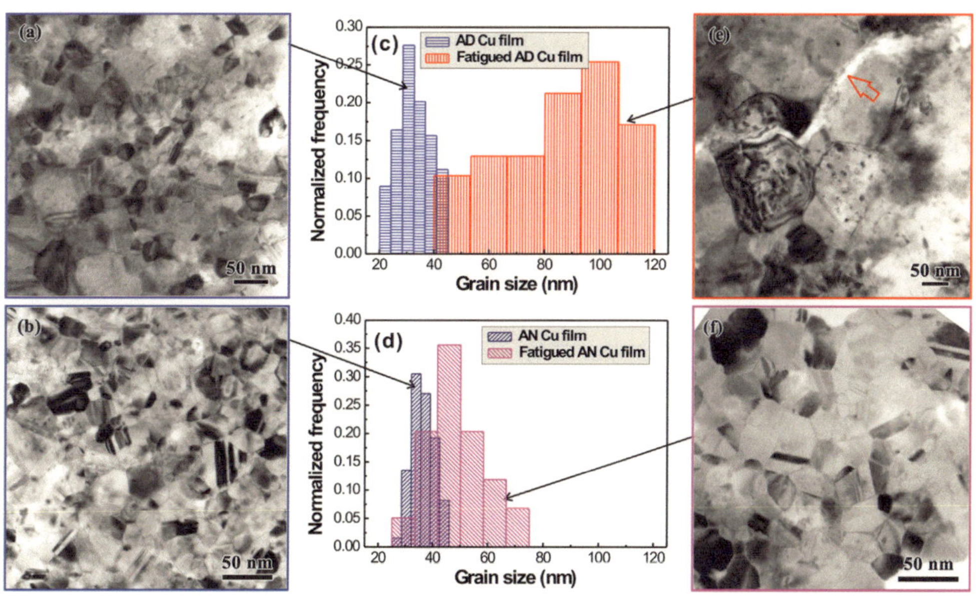

Figure 5. Microstructure of Cu films (a) as-deposited (AD) and (b) annealed (AN). (c and d) Statistical results of the grain size of the AD and the AN Cu films before and after fatigue test. Microstructure of Cu films after fatigue test (e) AD and (f) AN [49].

the mean grain size of both the as-deposited and annealed Cu samples increase after the fatigue test, as shown in **Figure 5**. However, the grain growth is greatly suppressed in the annealed NC Cu film, which leads to the enhanced resistance to the fatigue cracking as compared with that of the as-deposited one. The enhanced fatigue strength of Cu film after annealing may be related to the GB structure readjustment during the annealing process.

3.3. Grain boundary migration-theoretical model

Although the mechanical grain growth has been experimentally observed under various mechanical conditions, the underlying microstructural and atomic scale mechanisms are still open for debate. As mentioned in the introduction section, in an NC material, the fraction of atom residing at the GBs increases as the grain size decreases. Thus, GBs in NC materials promote the total free energy of the system. The reduction of this excess free energy through the removal of grain boundary area represents a large driving force for the grain growth. Grain growth in NC materials can be due to the rotation and coalescence of adjacent grains, as well as normal grain boundary movements. There are several theoretical models that have been proposed to describe the mechanically driven grain growth behavior, for example, stress-coupled GB migration [51–56] and grain rotation-induced grain coalescence [57].

The stress-coupled GB migration model is based on the argument that shear stress causes tangential movement of grains along GBs (GB sliding), and this produces a coupling with the normal motion of GBs (GB migration) [51]. Gutkin and Ovid'ko proposed a continuum disclination model for describing the stress-induced cooperative migration of an arbitrary tile GB [53–55]. The migrating GB was approximated by partial wedge disclination that can move under the applied shear stress, as shown in **Figure 6**. In the initial state, these GBs form two triple junctions. Under an applied shear stress, migration of GB3 from their initial position AB to a new position A′B′ occurs. Stress-induced migration of low-angle tilt GB3 results in the formation of two new triple junctions, A′ and B′. Straight-line defects (junctions) A, B, B′, and A′ are characterized by the disclination strength $\pm\omega$. The motion of the disclination produces rotational plastic deformation. The same is true for migration of two high-angle tile GBs with large angle gaps. It was shown that there is two critical stress, τ_{c1} and τ_{c2}, that controls the GB migration behavior. When the applied stress τ reaches τ_{c1}, the GB can migrate in the stable mode and their equilibrium position is determined by the level of τ. When $\tau > \tau_{c2}$, the GB migration becomes unstable when the GB propagation does not depend on the level of τ. In all cases, GB migration leads to the unstable growth of a grain at the expense of its neighbors. Energy methods calculation indicates that critical shear stresses strongly depend on the elastic modulus of the material, as well as on the strength of disclination-like defects appearing at the GB junctions in the process of GB migration.

For the case where the translational mode is mainly represented by GB sliding, Wang et al. suggested a theoretical model which describes the cooperative action of GB sliding and grain rotational deformation in mechanically loaded NC materials, as shown in **Figure 7** [25]. The grain rotation-induced grain coalescence model can be understood as follows [57]: with the applied force, GB dislocations glide results in the relative translational motion of GBs. However, the triple junctions impede the motion of GB dislocation. The blocked GB

http://dx.doi.org/10.5772/66426

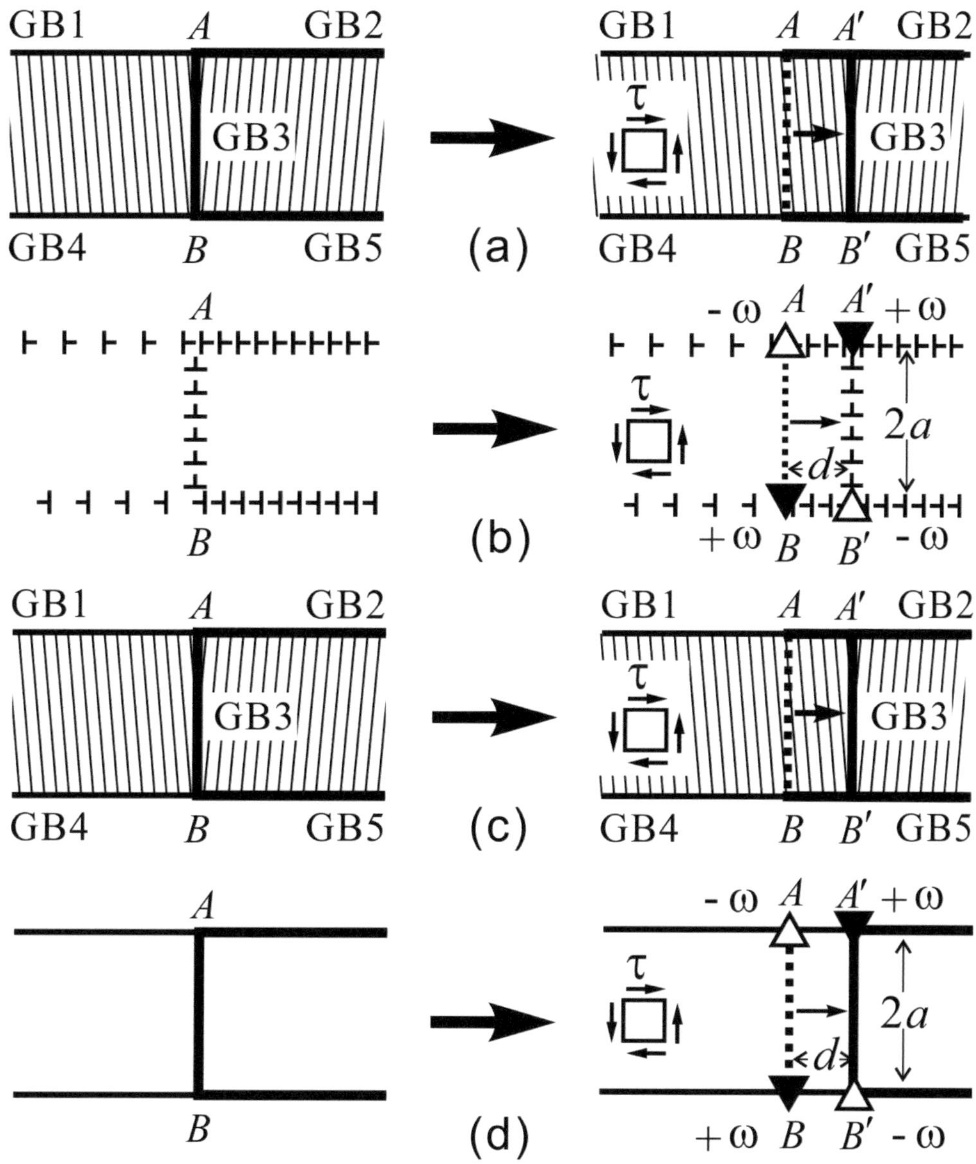

Figure 6. Schematic of stress-induced grain boundary (GB) migration (a and b) low-angle and (c and d) high-angle GB (GB3) by grain rotation through the glide of lattice dislocations (b) or motion of a dipole of wedge disclinations (d), respectively [54].

dislocation dissociates from the two climbing GB dislocation at the triple junction. With further plastic deformation, the dislocation split process happens repeatedly and the climbing GB dislocations form two dislocation walls along the GBs, which results in the rotation of the central NC grain. Multiple rotations bring the orientation of abutting grains closer together and reduce the GB misorientation angles, and even eliminate the GBs, leading to coalescence of smaller grains into larger ones.

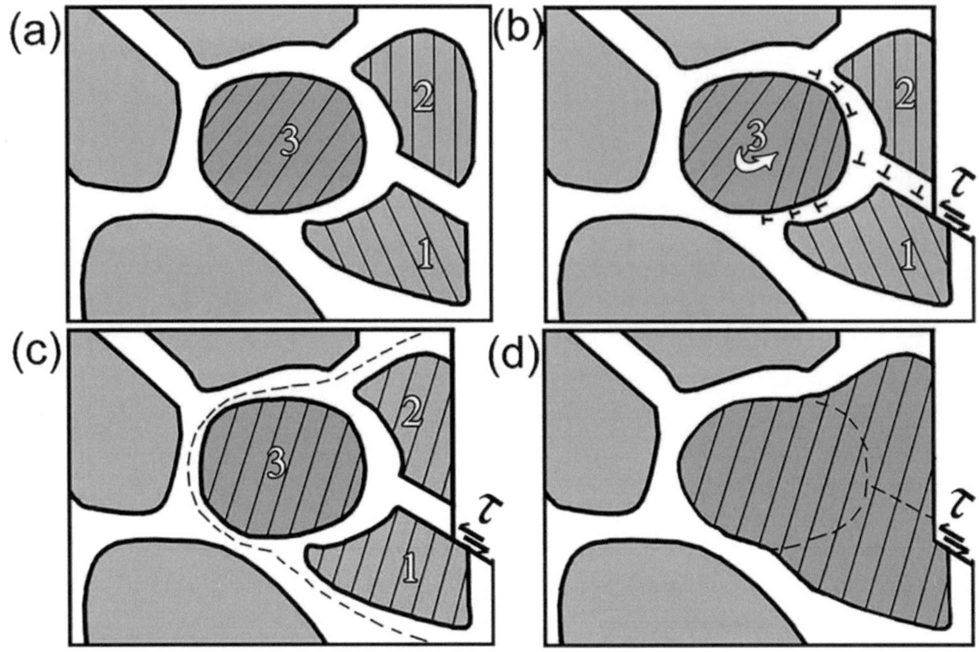

Figure 7. Combined grain boundary (GB) sliding and rotational deformation mode. (a) The GB of the nanocrystalline (NC) Ni dominated by high-angle GBs. (b) dislocations motion along GB impeded by the triple junction resulted in climbing GB dislocations, which caused the rotation of grain 3. (c) Repeated grain rotations leading to grain coalescence. (d) The formation of a larger grain with subGBs (highlighted by dotted line) [25].

3.4. Grain boundary migration-atomic scale mechanisms

Although theoretical models give some clues about the grain growth processes during mechanical deformation of NC materials, the exact microstructural scale and atomic scale mechanisms of the GB migration is still unclear. Quantitative information aimed to identify atomic scale mechanisms that reveal the influences of GB structure on the GB migration can be obtained by using in situ HRTEM and molecular dynamics (MD) simulation methods. Haslam et al. explored how grain rotation can induce grain coarsening [58]. They found that grain rotation decomposes a GB into multiple but distinct dislocations, which can then move by dislocation slip; these dislocations can annihilate at GBs or remain embedded within grains if the applied stress is relieved. The grain rotation mechanism purports that only some grains are able to rotate and thus coarsened grains can maintain their outer boundaries. This is consistent with the experimental result that GB of coarsened grain is characterized by low-angle boundaries [43, 44]. MD simulation of nanoindentation of NC Al films with a mean grain size of 7 nm showed that the grain rotation may be competing with the GB migration and the GB migration is likely dominant, as shown in **Figure 8** [59]. During cyclic deformation of 20-nm-thick Au thin film, Luo et al. found that grain growth of NC Au is closely correlated with twin formation [60]. Based on the atomic scale observations, as shown in **Figure 9**, they revealed that the formation of nanotwins is an effective way to assist grain coarsening. The grain coarsening process can be described as follows: the mutual nucleation of nanotwins near the GB changes the local

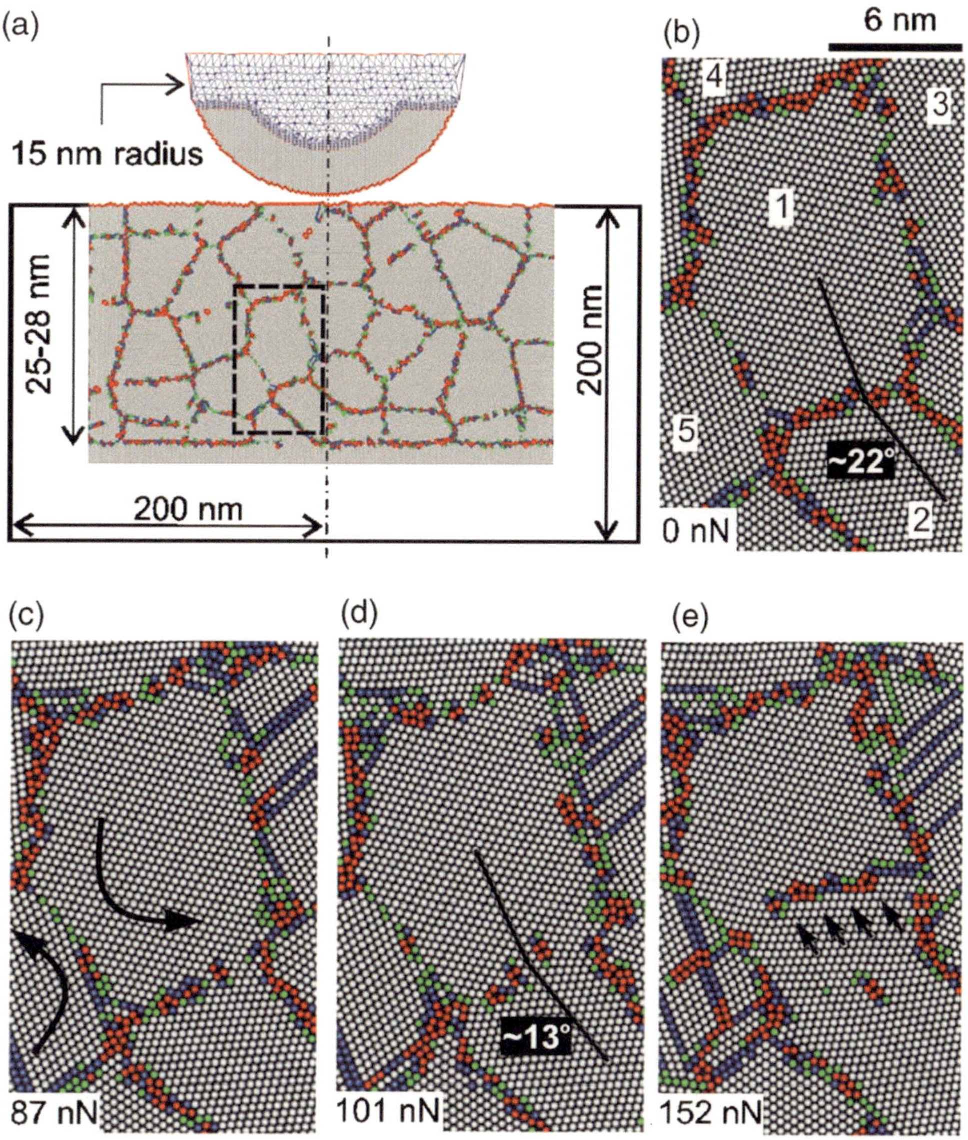

Figure 8. Atomic-level detail of the coalescence of grains. Deformation of Al film with grain size of 7 nm. (a) Contact zone before indentation simulation. (b–d) Rotation of grain 1 due to grain boundary (GB) sliding and transformation of the structure of the GBs. (e) Dissociation and migration of the GB between grains 1 and 2 [59].

grain misorientation and results in the GB dissociated into smaller segments, which is more mobile than their parent one.

Recently, by using in situ HRTEM, Luo et al. have reported the adjustment of GB structures of Cu film during a self-driven GB migration, which involves GB dissociation, partial dislocation emission from GB, and faceting/defaceting [61]. Furthermore, they revealed that GB migration ability is closely related to the local GB segment consisting of "hybrid" structural units.

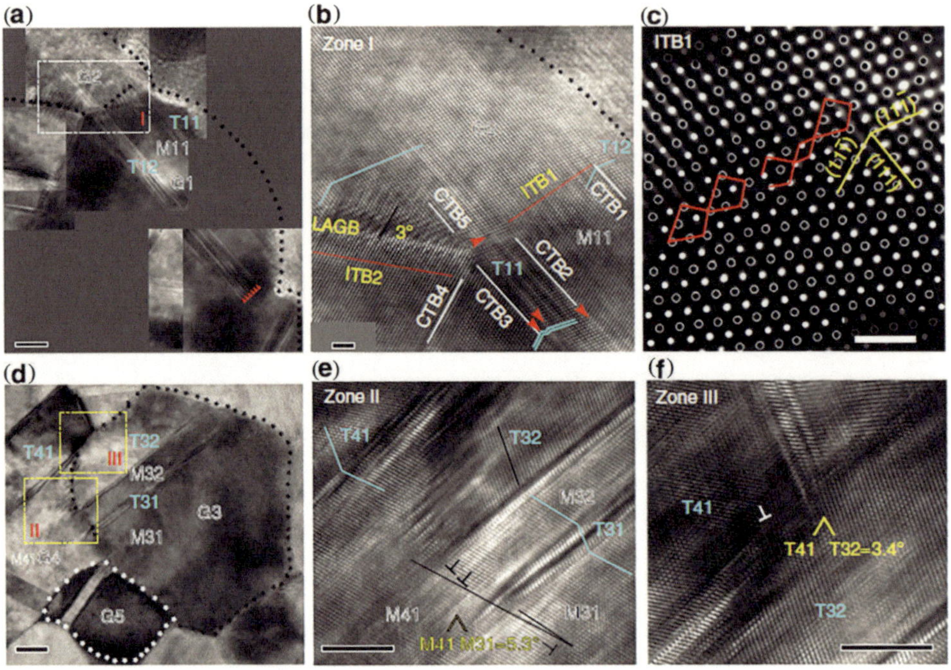

Figure 9. GB dissociation induced by twin formation. (a and d) TEM images of two typical examples of large grains with parallel multi-twins in the fatigued samples. (b and c) and (e and f) The corresponding details of the microstructures, respectively. Scale bars: (a and d–f) 5 nm, (b and c) 1 nm [60].

4. Summary and future work

As mentioned in the introduction section, the high volume fraction of atoms resided at GB makes the microstructural stability problem intrinsic to NC materials. In general, two approaches are used to stabilize the grain structure: by kinetically hindering the GB mobility or by thermodynamically lowering the GB energy through solute segregation [62]. Studies have shown that providing a short annealing treatment to allow for grain boundary relaxation can increase the fatigue life of Cu films [49]. HRTEM investigation on the detailed GB characters pre- and post-annealing processes will give more clues about the underlying mechanism. Recently, methods for stabilizing NC materials by the control of interface structure was reviewed by Lu [63]. It has been argued that nanostructures with a high-density of coherent twin boundaries (CTB), which are low energy, low mobility boundaries with a high degree of crystallographic ordering, would provide the required resistance to the thermal coarsening while enhancing the strength. The addition of solute atoms (alloying) can pin the GB and/or lower the GB energy, improving the microstructural stability of NC [64, 65]. However, due to the complexity induced by alloying, more experimental and theoretical investigations on the structural characteristics need to be done in the future.

Acknowledgements

This research was supported by the Australian Research Council Discovery Project (DP150101717) and the Joint Foundation of Liaoning Province National Science Foundation and Shenyang National Laboratory for Materials Science (2015021005).

Author details

Xiaofei Zhu[1, 2], Guangping Zhang[2] and Cheng Yan[1, *]

*Address all correspondence to: c2.yan@qut.edu.au

1 School of Chemistry, Physics and Mechanical Engineering, Science and Engineering Faculty, Queensland University of Technology (QUT), Brisbane, Queensland, Australia

2 Shenyang National Laboratory for Materials Science, Institute of Metal Research, Chinese Academy of Sciences (CAS), Shenyang, China

References

[1] Beyerlein IJ, Demkowicz MJ, Misra A, Uberuaga BP. Defect-interface interactions. Prog Mater Sci. 2015;74:125–210.

[2] Hirth JP, Wang J, Tomé CN. Disconnections and other defects associated with twin interfaces. Prog Mater Sci. 2016;83:417–71.

[3] Kumar KS, Van Swygenhoven H, Suresh S. Mechanical behavior of nanocrystalline metals and alloys. Acta Mater. 2003;51(19):5743–74.

[4] Meyers MA, Mishra A, Benson DJ. Mechanical properties of nanocrystalline materials. Prog Mater Sci. 2006;51(4):427–556.

[5] Fultz B, Frase HN. Grain boundaries of nanocrystalline materials – their widths, compositions, and internal structures. Hyperfine Interact. 2000;130(1):81–108.

[6] Spearot DE, McDowell DL. Atomistic modeling of grain boundaries and dislocation processes in metallic polycrystalline materials. J Eng Mater Technol. 2009;131(4):041204.

[7] Siegel RW, Thomas GJ. Grain boundaries in nanophase materials. Ultramicroscopy. 1992;40(3):376–84.

[8] Suryanarayana C. The structure and properties of nanocrystalline materials: issues and concerns. JOM. 2002;54(9):24–7.

[9] Frøseth AG, Derlet PM, Van Swygenhoven H. Dislocations emitted from nanocrystalline grain boundaries: nucleation and splitting distance. Acta Mater. 2004;52(20):5863–70.

[10] LaGrange T, Reed BW, Wall M, Mason J, Barbee T, Kumar M. Topological view of the thermal stability of nanotwinned copper. Appl Phys Lett. 2013;102(1):011905.

[11] Hasnaoui A, Van Swygenhoven H, Derlet PM. Dimples on nanocrystalline fracture surfaces as evidence for shear plane formation. Science. 2003;300(5625):1550–2.

[12] Watanabe T. Grain boundary engineering: historical perspective and future prospects. J Mater Sci. 2011;46(12):4095–115.

[13] Brandon DG. The structure of high-angle grain boundaries. Acta Metall. 1966;14(11): 1479–84.

[14] Read WT, Shockley W. Dislocation models of crystal grain boundaries. Phys Rev. 1950;78(3):275–89.

[15] Sutton AP, Balluffi RW. On geometric criteria for low interfacial energy. Acta Metall. 1987;35(9):2177–201.

[16] Tschopp MA, Solanki KN, Gao F, Sun X, Khaleel MA, Horstemeyer MF. Probing grain boundary sink strength at the nanoscale: energetics and length scales of vacancy and interstitial absorption by grain boundaries in α-Fe. Phys Rev B. 2012;85(6):064108.

[17] Bishop GH, Chalmers B. A coincidence-ledge-dislocation description of grain boundaries. Scripta Metall. 1968;2(2):133–9.

[18] Rittner JD, Seidman DN. <110> symmetric tilt grain-boundary structures in fcc metals with low stacking-fault energies. Phys Rev B. 1996;54(10):6999–7015.

[19] Oba F, Ohta H, Sato Y, Hosono H, Yamamoto T, Ikuhara Y. Atomic structure of [0 0 0 1]-tilt grain boundaries in ZnO: a high-resolution TEM study of fiber-textured thin films. Phys Rev B. 2004;70(12):125415.

[20] Zhang K, Weertman JR, Eastman JA. The influence of time, temperature, and grain size on indentation creep in high-purity nanocrystalline and ultrafine grain copper. Appl Phys Lett. 2004;85(22):5197–9.

[21] Soer WA, Hosson JTMD, Minor AM, Morris Jr JW, Stach EA. Effects of solute Mg on grain boundary and dislocation dynamics during nanoindentation of Al–Mg thin films. Acta Mater. 2004;52(20):5783–90.

[22] Gianola DS, Van Petegem S, Legros M, Brandstetter S, Van Swygenhoven H, Hemker KJ. Stress-assisted discontinuous grain growth and its effect on the deformation behavior of nanocrystalline aluminum thin films. Acta Mater. 2006;54(8):2253–63.

[23] Pan D, Kuwano S, Fujita T, Chen MW. Ultra-large room-temperature compressive plasticity of a nanocrystalline metal. Nano Lett. 2007;7(7):2108–11.

http://dx.doi.org/10.5772/66426

[24] Rupert TJ, Gianola DS, Gan Y, Hemker KJ. Experimental observations of stress-driven grain boundary migration. Science. 2009;326(5960):1686–90.

[25] Wang YB, Li BQ, Sui ML, Mao SX. Deformation-induced grain rotation and growth in nanocrystalline Ni. Appl Phys Lett. 2008;92(1):011903.

[26] Wang L, Teng J, Liu P, Hirata A, Ma E, Zhang Z, et al. Grain rotation mediated by grain boundary dislocations in nanocrystalline platinum. Nat Commun. 2014;5:4402.

[27] De Hosson JTM, Soer WA, Minor AM, Shan Z, Stach EA, Syed Asif SA, et al. In situ TEM nanoindentation and dislocation-grain boundary interactions: a tribute to David Brandon. J Mater Sci. 2006;41(23):7704–19.

[28] Zhang K, Weertman JR, Eastman JA. Rapid stress-driven grain coarsening in nanocrystalline Cu at ambient and cryogenic temperatures. Appl Phys Lett. 2005;87(6):061921.

[29] Jin M, Minor AM, Stach EA, Morris Jr JW. Direct observation of deformation-induced grain growth during the nanoindentation of ultrafine-grained Al at room temperature. Acta Mater. 2004;52(18):5381–7.

[30] Gai PL, Zhang K, Weertman J. Electron microscopy study of nanocrystalline copper deformed by a microhardness indenter. Scripta Mater. 2007;56(1):25–8.

[31] Su YH, Tomota Y, Harjo S, Adachi Y. Deformation-induced grain coalescence in an electrodeposited pure iron sheet studied by in situ neutron diffraction and electron backscatter diffraction. Acta Mater. 2012;60(8):3393–401.

[32] Fan GJ, Fu LF, Choo H, Liaw PK, Browning ND. Uniaxial tensile plastic deformation and grain growth of bulk nanocrystalline alloys. Acta Mater. 2006;54(18):4781–92.

[33] Fan GJ, Fu LF, Qiao DC, Choo H, Liaw PK, Browning ND. Grain growth in a bulk nanocrystalline Co alloy during tensile plastic deformation. Scripta Mater. 2006;54(12):2137–41.

[34] Gertsman VY, Birringer R. On the room-temperature grain growth in nanocrystalline copper. Scripta Metall Mater. 1994;30(5):577–81.

[35] Ames M, Markmann J, Karos R, Michels A, Tschöpe A, Birringer R. Unraveling the nature of room temperature grain growth in nanocrystalline materials. Acta Mater. 2008;56(16):4255–66.

[36] Günther B, Kumpmann A, Kunze HD. Secondary recrystallization effects in nanostructured elemental metals. Scripta Metall Mater. 1992;27(7):833–8.

[37] Gianola DS, Mendis BG, Cheng XM, Hemker KJ. Grain-size stabilization by impurities and effect on stress-coupled grain growth in nanocrystalline Al thin films. Mater Sci Eng A. 2008;483–484:637–40.

[38] Witney AB, Sanders PG, Weertman JR, Eastman JA. Fatigue of nanocrystalline copper. Scripta Metall Mater. 1995;33(12):2025–30.

[39] Zhang K, Weertman JR. Microstructural changes produced by fatigue in high-purity copper severely deformed by cryogenic rolling. Metall Mater Trans A. 2009;40(10):2255–63.

[40] Padilla HA, Boyce BL. A review of fatigue behavior in nanocrystalline metals. Exp Mech. 2010;50(1):5–23.

[41] Cheng S, Zhao Y, Wang Y, Li Y, Wang X-L, Liaw PK, et al. Structure modulation driven by cyclic deformation in nanocrystalline NiFe. Phys Rev Lett. 2010;104(25):255501.

[42] Boyce BL, Padilla HA. Anomalous fatigue behavior and fatigue-induced grain growth in nanocrystalline nickel alloys. Metall Mater Trans A. 2011;42(7):1793–804.

[43] Meirom RA, Alsem DH, Romasco AL, Clark T, Polcawich RG, Pulskamp JS, et al. Fatigue-induced grain coarsening in nanocrystalline platinum films. Acta Mater. 2011;59(3):1141–9.

[44] Meirom RA, Clark TE, Muhlstein CL. The role of specimen thickness in the fracture toughness and fatigue crack growth resistance of nanocrystalline platinum films. Acta Mater. 2012;60(3):1408–17.

[45] Kwan CCF, Wang Z. Strain incompatibility and its influence on grain coarsening during cyclic deformation of ARB copper. Philos Mag. 2012;93(9):1065–79.

[46] Park H-K, Kang H-G, Park C-S, Huh M-Y, Hwang N-M. Ex situ observation of microstructure evolution during abnormal grain growth in aluminum alloy. Metall Mater Trans A. 2012;43(13):5218–23.

[47] Greene RB, Fünfschilling S, Fett T, Hoffmann MJ, Kruzic JJ, Roedel J. Fatigue crack growth behavior of silicon nitride: roles of grain aspect ratio and intergranular film composition. J Am Ceram Soc. 2013;96(1):259–65.

[48] Horky J, Khatibi G, Setman D, Weiss B, Zehetbauer MJ. Effect of microstructural stability on fatigue crack growth behaviour of nanostructured Cu. Mech Mater. 2013;67:38–45.

[49] Zhang B, Xiao TY, Luo XM, Zhu XF, Zhang GP. Enhancing fatigue cracking resistance of nanocrystalline Cu films on a flexible substrate. Mater Sci Eng A. 2015;627:61–4.

[50] Bufford DC, Stauffer D, Mook WM, Syed Asif SA, Boyce BL, Hattar K. High cycle fatigue in the transmission electron microscope. Nano Lett. 2016;16(8):4946–53.

[51] Cahn JW, Mishin Y, Suzuki A. Coupling grain boundary motion to shear deformation. Acta Mater. 2006;54(19):4953–75.

[52] Cahn JW, Taylor JE. A unified approach to motion of grain boundaries, relative tangential translation along grain boundaries, and grain rotation. Acta Mater. 2004;52(16):4887–98.

[53] Gutkin MY, Mikaelyan KN, Ovid'ko IA. Athermal grain growth through cooperative migration of grain boundaries in deformed nanomaterials. Scripta Mater. 2008;58(10):850–3.

[54] Gutkin MY, Ovid'ko IA. Grain boundary migration as rotational deformation mode in nanocrystalline materials. Appl Phys Lett. 2005;87(25):251916.

[55] Bobylev SV, Gutkin MY, Ovid'ko IA. Transformations of grain boundaries in deformed nanocrystalline materials. Acta Mater. 2004;52(13):3793–805.

[56] Sansoz F, Molinari JF. Mechanical behavior of Σ tilt grain boundaries in nanoscale Cu and Al: A quasicontinuum study. Acta Mater. 2005;53(7):1931–44.

[57] Gutkin MY, Ovid'ko IA, Skiba NV. Crossover from grain boundary sliding to rotational deformation in nanocrystalline materials. Acta Mater. 2003;51(14):4059–71.

[58] Haslam AJ, Moldovan D, Yamakov V, Wolf D, Phillpot SR, Gleiter H. Stress-enhanced grain growth in a nanocrystalline material by molecular-dynamics simulation. Acta Mater. 2003;51(7):2097–112.

[59] Sansoz F, Dupont V. Grain growth behavior at absolute zero during nanocrystalline metal indentation. Appl Phys Lett. 2006;89(11):111901.

[60] Luo XM, Zhu XF, Zhang GP. Nanotwin-assisted grain growth in nanocrystalline gold films under cyclic loading. Nat commun. 2014;5:3021.

[61] Luo XM, Zhang B, Zhu XF, Zhou YT, Xiao TY, Zhang GP. Local-structure-affected behavior during self-driven grain boundary migration. MRS Commun. 2016;6(02):85–91.

[62] Tschopp MA, Murdoch HA, Kecskes LJ, Darling KA. “Bulk” nanocrystalline metals: review of the current state of the art and future opportunities for copper and copper alloys. JOM. 2014;66(6):1000–19.

[63] Lu K. Stabilizing nanostructures in metals using grain and twin boundary architectures. Nature Rev Mater. 2016;1:16019.

[64] Koch CC, Scattergood RO, Darling KA, Semones JE. Stabilization of nanocrystalline grain sizes by solute additions. J Mater Sci. 2008;43(23):7264–72.

[65] Chookajorn T, Murdoch HA, Schuh CA. Design of stable nanocrystalline alloys. Science. 2012;337(6097):951–4.

Chapter 8

Control the Metal Grain Boundary Recrystallization Evolution by the Laser Radiation Electric Field Strength Direction Under Cyclic Thermal Loading

Makin Vladimir Sergeevich

Additional information is available at the end of the chapter

http://dx.doi.org/10.5772/66248

Abstract

The spatial grating formation at metal surface under the linear polarized laser radiation action is briefly considered. The spatial grating periods are well described in framework of universal polariton model (UPM) and are well-defined physical quantities. The production of new-type gratings (quasi-gratings) at laser power densities lower than the metal melting threshold with power-dependent typical spatial scale and polarization-dependent orientation are discovered. The regularities of quasi-grating production are experimentally studied. The physical model of quasi-grating formation explaining the anisotropic character of metal recrystallization is suggested. The anisotropy is caused by the directed electron flux interaction with grain boundaries. The electron flux results from the drag effect of electrons by surface plasmon polaritons (SPPs). SPPs are excited by incident laser radiation on the surface irregularities including the grain boundaries. The volume analog of considered effect is the electroplastic one, and some of its regularities are considered.

Keywords: metal, anisotropic recrystallization, laser radiation, linear polarization, grain boundary movement, surface plasmon polaritons, electrons drag by surface plasmon polaritons, electroplastic effect

1. Introduction

It is known that the separated direction usually arises on the condensed matter surfaces and in bulk under the action of linear polarized laser radiation. The arising phenomena of linear polarized electric field strength-oriented grating formation are well described in framework of universal polariton model (UPM) of laser-induced condensed matter damage [1]. The UPM well describes the spatial periods for normally oriented (**g** | | **E**) [2] and abnormally oriented (**g** ⊥ **E**) gratings (**g**) as for long pulse durations [3] as for ultrashort laser pulses [4, 5]. Here **E** is the electric field strength vector of incident laser radiation. In the later case, the effect exists for condensed media with different physical properties: metals, semiconductors, and dielectrics. So the peculiar directions arise due to the vector nature of light, surface plasmon polaritons [6] and channel (wedge) surface plasmon polariton [4] excitation, and participation in the interference process. The produced spatial gratings have some distribution in directions and periods for ultrashort pulse durations, but for long pulses depending on laser wavelength and optical properties of boundary materials, the periods have well-defined values. In our experiments with the incident laser radiation of nanosecond duration, the field-oriented grating is formed with periods defined not so strictly as the universal polariton model dictated. So, the problem is the physical mechanism of the observed effect.

The contemporary theory of condensed media describes the spatially hierarchical synergetic behavior of structures in solids, including under conditions of relaxation from nonequilibrium state [7]. One example is the metal recrystallization [8], including laser-induced recrystallization [9]. It is known that the process of collecting recrystallization on metal's surface caused by heating up to the melting point is followed by grain boundary movement and enlarging of their scales [10]. This is the isotropic process because of the lack of separated direction in considered system. There are known experiments on metal films deposited on dielectric substrate recrystallization under cyclic heating by laser radiation up to the melting point followed by lateral spatial grain-scale enlargement [9]. The grain-scale growth in average is isotropic due to the absence of separated direction in considered system *metal-dielectric-laser radiation*.

2. Results

We studied the self-organized phenomena of micro- and nanostructure formation on metal surfaces under action of series of laser radiation pulses (λ = 1064 μm, τ = 10, 100 ns) at near-normal incidence upon the mechanically polished titanium surface (VT—1-0). Formation of three types of structures versus the laser power density (q) and pulse number (N) was observed. At q values corresponding to partial metal melting during the pulse, the creation of resonant periodic grating with period

$$d_0 = \lambda / \eta \approx 1.04 \ \mu m \tag{1}$$

and orientation $\mathbf{g_0}$ | | **E** written as surface relief modulation was observed. Here η is the real part of the refractive index for titanium-air boundary for surface plasmon polaritons. Their formation was due to the interference of the incident laser radiation and surface plasmon polaritons excited by it (universal polariton model (UPM) [11]).

The second type of the structures (abnormal one) was also observed in the regime of partial metal melting; it was observed predominantly on the ridges of the main resonant relief (of the first-type gratings) and had the period [5]:

$$d = \lambda / 4\xi \approx 0.21 \ \mu m \tag{2}$$

and anomalous orientation $\mathbf{g} \perp \mathbf{E}$. Their formation was caused by participation of the wedge surface plasmon polaritons (WSPPs) guided by ridges of the main resonant relief (structures of the first type) explained by the nonlinear mathematical model of spatial period formation [12]. Here ξ is the real part of the refractive index for wedge surface plasmon polaritons of the considered boundary.

The third type of structures was observed only for laser pulse duration $\tau \approx 10$ ns at q values not exceeding the material melting threshold. The produced microrelief was the quasi-grating **G**, the period of which varied by laser radiation power density and orientation was $\mathbf{G} \perp \mathbf{E}$. Note that in contrast to the resonant micro- and nanostructures (of first and second types), the quasi-gratings had no precise value of the period according to formula (1) or (2). So, the third type of structures could not be explained in the framework of the universal polariton model. To explain their occurrence, the original model was developed.

As the samples in experiments the VT1-0 titanium plates mechanically polished with optical quality were used (geometrical sizes: 7 mm diameter and 1 mm thickness). The Q-switched linear polarized radiation of Nd^{3+}:Yttrium Aluminum Garnet (YAG) laser ($\lambda = 1064$ nm, $\tau = 10$ ns, 100 ns, $f \leq 12.5$ Hz) with power density $q \leq 0.5$ MW/cm^2 was used for sample irradiation. Laser radiation was focused by lens with focal length f = 18 cm into the typical diameter of irradiated spot (0.6 ÷ 1) mm. The sample surface reflectivity dynamics from pulse to pulse at $\lambda = 632.8$ nm was measured with the help of integrated sphere. The residual surface relief was studied by optical microscopy (preferentially in dark field) and atomic force microscopy. The main experiments were made with normal incidence of laser radiation or for p-polarized radiation and not high angles of incidence.

In experiments, the following evolution of the micro- and nanorelief was observed, which can be by convention divided by three consecutive stages. The *first stage* was the fine-scale formation of quasi-isotropic nanorelief, the typical size of which was less than the optical microscope resolution value in mode of fine-scale deformation grain boundary network. The *second stage* was the rise of mean grain-size dimension with the grain anisotropy appearance. The grain boundaries were observable with optical microscopy. This is the typical stage of collective recrystallization. At the *third stage*, the grain boundary was gradually converted into quasi-linear relief having typical size $s \sim (4 \div 6)$ μm. The relief was formed in mode of thermal grooves (see **Figure 1**). Its quasi-grating vector **G** was near perpendicular to $\mathbf{E}_t$: Here $\mathbf{E}_t$ is the tangential projection of electric field strength vector of the incident radiation. Especially note that the *s* value was power density dependent and rising with *q*. The *s* value also was varied from the center to periphery of irradiated spot (see **Figures 2** and **3**). This dependence was the indicator of that the structure appearance is not in the framework of UPM. After the action of approximately N ≥ 300 pulses at the periphery of irradiated zone, the low contrast resonant surface relief with period $d^- = \lambda/(\eta - \sin\theta)$ ($\mathbf{g}^- \parallel \mathbf{E}_t$) was observed for <30°. Here θ is the angle of incidence of laser radiation. For >30° the quasi-grating **G** disappears, and only the grating $\mathbf{g}^-$ appears at the central part of spot expanding toward periphery frequently in mode of separated tracks which fill much more area with angle θ. Note here that the effective sources of SPP's scattering and incident laser radiation transformation into SPPs are grain boundaries (see, for instance, Ref. [13]).

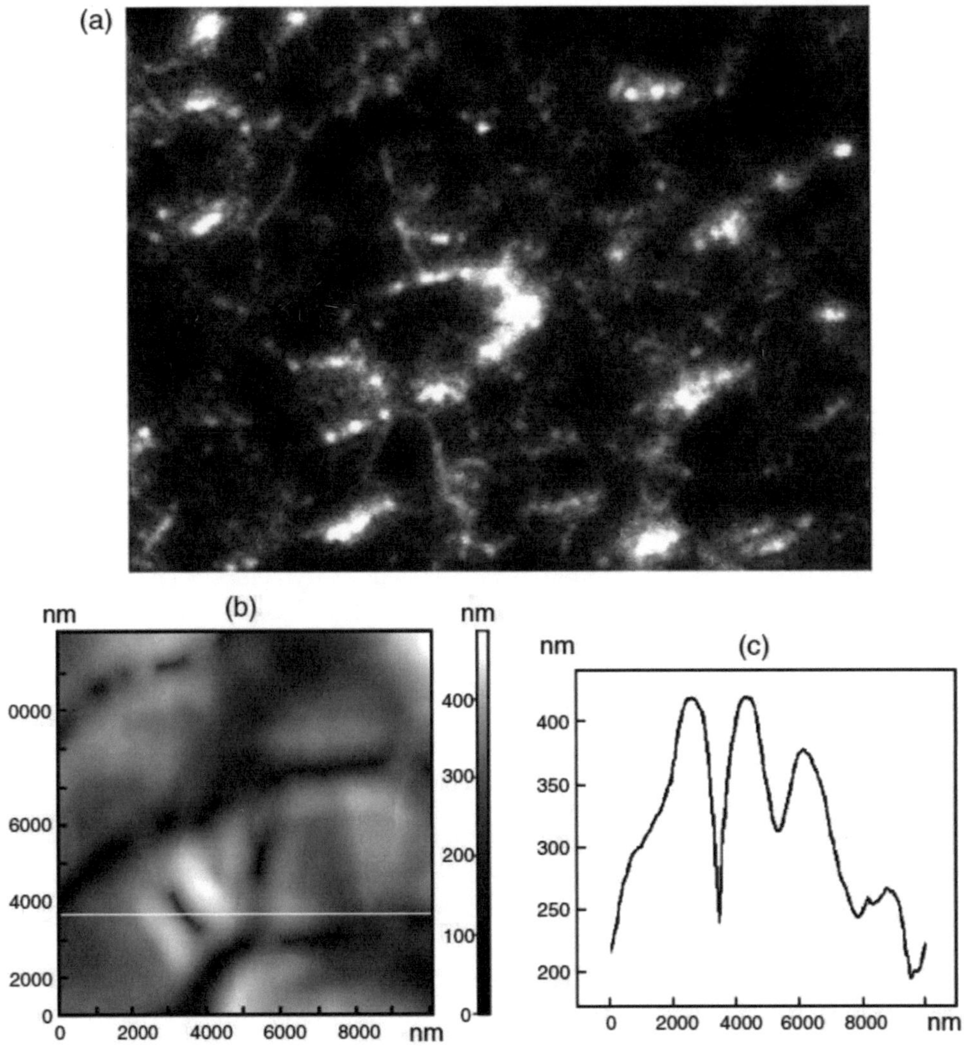

Figure 1. Topography of titanium surface produced under the interaction of 60 pulses of linear polarized laser radiation with power density ≈0.8 MW/cm²: (a) the initial stage of anisotropic grain growth dark-field optical microtopography, (b) the enlarged view of atomic force topography, and (c) typical profilogram of irradiated surface made along the white line indicated in (b).

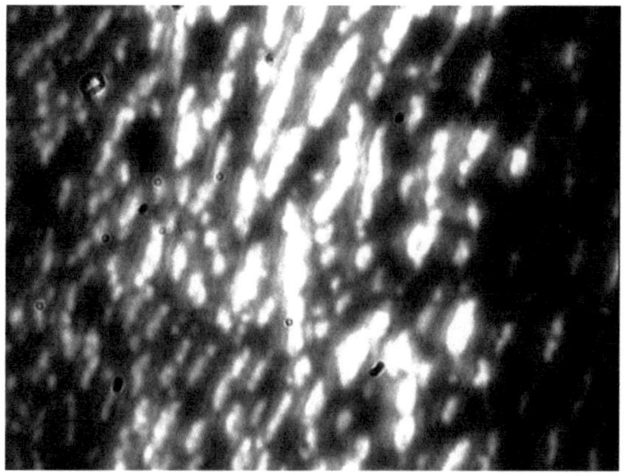

Figure 2. Dark-field optical microscopy image of the titanium surface spot area irradiated by linear polarized laser radiation (λ = 1064 nm, τ = 20 ns) shows the quasi-grating formation of grooves of thermal grooving.

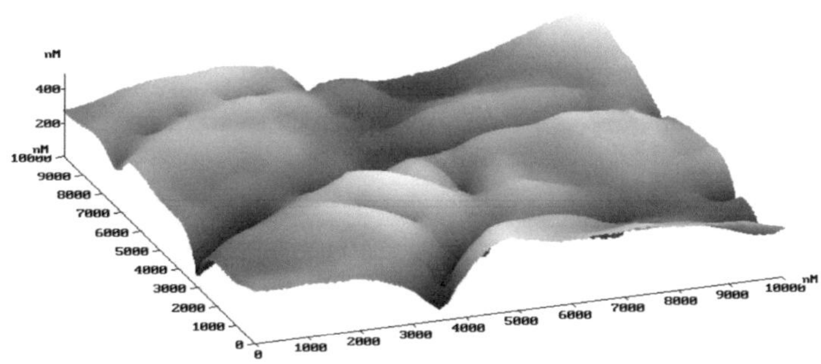

Figure 3. Image of titanium surface irradiated by series of N = 62 pulses of linear polarized laser radiation with $q < q_{melt}$ obtained by atomic force microscopy.

The experiments also were conducted in atmosphere of active and inert gases at atmosphere pressure. The number of laser pulses needed to produce relief of given height in inert gases was higher by (1.5 ÷ 2) times in comparison with air and was lower by (2 ÷ 2.5) times in oxygen atmosphere. As is known the formation of resonant gratings by circular polarized laser radiation is difficult because the gratings of all possible orientations (but discrete) must be produced. For this case, the degree of positive feedback via grating height occurs to be insufficient. So, in our experiments for circular polarization, neither resonant gratings nor quasi-grating formation was observed. The experiments were made at laser power density in the range (0.3 ÷ 0.8) MW/cm^2 to be sure that the melting point of titanium surface will not be achieved.

To study pulse-to-pulse reflectivity dynamics of irradiated area, the integrated sphere and probe radiation (λ = 632.8 nm) were used. The time dependence of surface absorptivity has shown the shallow minima for N ≈ 40 followed by gradual rise. The absorptivity change at minima was $\Delta A = A_0 - A_{min} \leq 0.14$, where A_0 is the initial absorptivity value and the A_{min} value corresponds to surface absorptivity minima; the initial value is $A_0 = 0.5$ (see, for instance, Ref. [14]). In our experiments, the final value of A was near 1.2 A_0.

The dynamics of surface relief changes was caused by grain boundary displacement in the surface layer of the order of the depth of surface layer pulsed heating. The local minima existence in the function A = A(N) is the consequence of two following process competitions, namely, the collective recrystallization which enhances the optical properties and causes the metal absorptivity falling and oxygen dissolution in metal skin layer and surface oxide film growth cause the absorptivity rising. The experimentally observed evolution of the surface relief on the initial stage is the consequence of technique of its polishing. Really, in the process of mechanical polishing, the metastable highly cold-hardening layer of titanium is formed having properties approaching to ones of amorphous metal. The action of repetitive pulses of laser radiation leads to more equilibrium metal state through the recrystallization process [10]. So in a whole, the process is followed by sufficient grain-size growth of surface layer.

Next discuss the quasi-grating vector **G** orientation correlated with the laser radiation polarization origin. The propagation directions of excited SPPs are mainly along the E_t vector. That is why SPPs most efficiently interact with grain boundaries which have orthogonal to $\mathbf{E}_t$ orientation. The directional SPP's propagation causes the appearance of current of electrons in metal skin layer [15, 16]. The current appears due to the tangential component of the Lorentz force $F_t \sim [v \times H]$ where v is the electron velocity and H is SPP's magnetic field strength vector. The action of the SPP's vertical component of electric field strength vector (E_z) accelerates the skin-layer electrons in vertical direction (v_z) which produces the Lorentz force component along SPP's propagation direction, $\mathbf{F}_t \| (k_s/k_s)$. The sign of vertical component of the electric field strength (E_z) is changing simultaneously with the sign of magnetic field strength (H) that is why the direction of the Lorentz force component $\mathbf{F}_t$ always remains along the SPP propagation direction k_s/k_s.

The directed flux of electrons interacts with grain boundaries as with a wall, thus supplying an additional force action and directed grain displacement. Obviously, maximal momentum will transfer to the deformation boundary, which is orthogonal to the SPP's propagation direction. In such a way, the anisotropy of grain growth is rising.

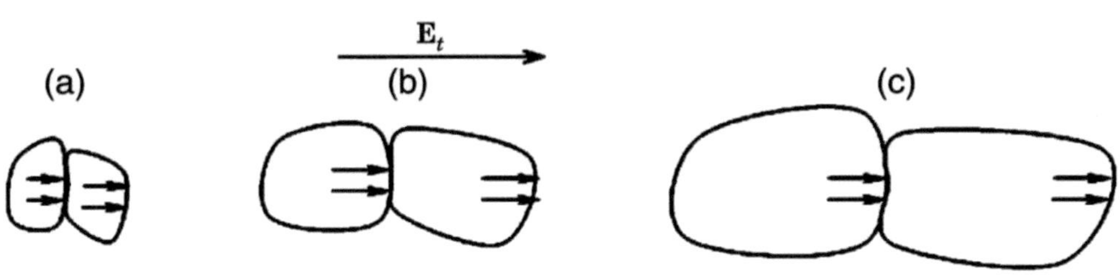

Figure 4. Scheme illustrated the subsequent grain boundary evolution under their interaction with directed flux of skin-layer electrons dragged by laser-excited surface plasmon polaritons. The progressive stages of the grain growth are shown in (a), (b), and (c). The vector orientation of quasi-grating formed at the final stage is $G \perp E_t$.

http://dx.doi.org/10.5772/66248

The process of quasi-grating formation occurs with positive feedback. Qualitatively, the directed flux of electrons action is mostly effective for the grain boundary orientation $\boldsymbol{n} \parallel \boldsymbol{k}_s$, where $\boldsymbol{n}$ is normal to the grain boundary laying in the plane of irradiated surface and $\boldsymbol{k}_s$ is the wave vector of SPP's. Taking in mind the grain boundary continuity, the neighbor boundary areas' curvature reduces. This enhance the efficiency of SPP's excitation (in given direction). This brings about the rise of Lorentz force Ft acting on grain boundaries of neighbor grains for which the direction of normal n is nearly parallel to $\boldsymbol{E}_t$, (**Figure 4**) This process takes place with the positive feedback and spreads over the irradiated zone. Finally, the self-consistent situation arises in which effective SPP's excitation is realized mainly on almost linear grain boundaries with $\boldsymbol{n} \parallel \boldsymbol{k}_s$. The process is stabilized when SPP's intensity is insufficient to move the grain boundary and their energy is dissipated into the metal. Such situation occurs for the optimal values of typical distances between the neighborhood grooves of formed quasi-gratings of the order of SPP's propagation length $L = 1/\alpha$, where α is the attenuation coefficient of the metal-air boundary for SPPs. The spatial areas of irradiated spot with higher laser radiation intensity are awaited to have the quasi-grating periods higher than for lower power density areas due to nearly linear dependence of excited SPP's intensities on the intensity of incident radiation. This conclusion is supported by our experimental results.

Let us estimate the value of SPP's attenuation coefficient α using the tabulated optical constants for $\lambda = 1064$ nm [14]:

$$\alpha = 2\,\mathrm{Im}k_s = 2k_0\left(\frac{\varepsilon_m}{\varepsilon_m + 1}\right)^{1/2}. \tag{3}$$

Here k_s is the SPP's wave vector's module, ε_m is titanium metal complex dielectric permittivity, $k_0 = \omega/c$ is the wave number of laser radiation in vacuum, ω is the circular frequency of laser radiation, and c is the velocity of light in vacuum. The estimate based on expression (3) and optical constants of titanium for $\lambda = 1064$ nm shows that the SPP's propagation length $L = \alpha^{-1} \approx 4\lambda$ for $\lambda = 1064$ nm and this value well coincides with experimentally measured value $s \approx 5$ μm. For surface areas with higher power density, the s value may reach (6 ÷ 6.5) μm (see **Figure 2**).

The driving force to displace the grain boundaries is the result of the flux of moving electrons dragged by SPP's interaction with grain boundaries, is limited in depth by titanium skin layer, and is of the order lower than the thermal heating depth. The skin-layer size spatial localization of driving force makes the boundaries displacement process easier. At the final stage of evolution, the $\boldsymbol{G}$ quasi-grating relief performs the nearly parallel and equidistant grooves of thermal etching (see **Figures 5a**, **5b**, and **1c**), having the inverse knife structure shapes with the depth up to 500 nm. The cross section of groove has radius of tip curvature of the order 50 nm or less, and dihedral angle at the tip is of the order 120°.

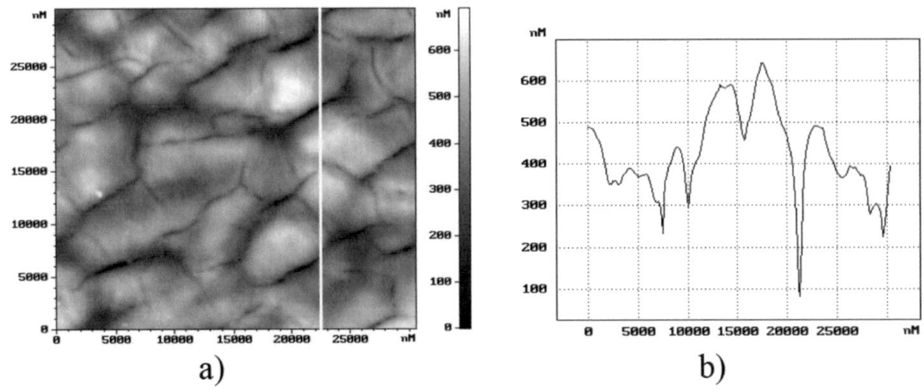

Figure 5. (a) Enlarged fragment of surface area of produced thermal grooves and (b) cross-sectional profile of surface obtained along white horizontal line of **Figure 4a**.

From **Figure 3**, it follows that as a result of multipulsed irradiation, the titanium surface locally becomes sufficiently more smooth than the initial one (surface roughness after surface mechanical polishing, stretching, and so on is smoothing; see **Figure 1b**). This is a well-known result of material redistribution caused by surface atom diffusion [17].

The inverse knife-shaped metal groove can support the channel surface plasmon polariton (CSPP) propagation [3, 5, 18]. In considered geometry of experiment, the direction of quasi-grating vector is orthogonal to laser radiation polarization E_t. For efficient excitation of CSPP, the electric field strength component of incident radiation must be maximal. So, the metal surface covered with quasi-grating has anisotropic absorptivity due to CSPP excitation and dissipation of their energy into heat. It is known that micro- and nanostructured metal surface also has anisotropy of electrical properties [19].

Note that the effect of the electrons drag by surface plasmon polaritons becomes apparent in the surface current in metal skin layer [12, 15] and in the *lateral* flux of relativistic electrons in vacuum under the metal surface irradiation by exawatt laser power density (pulse duration less than 1 ps) [20].

One may wait that the discovered effect may be well observable for powerful ultrashort laser pulse interaction with metals. Really, the experimental data for the multipulsed laser interaction with metals and alloys have been published for femtosecond pulse durations followed by quasi-grating $\mathbf{G} \perp \mathbf{E}$ formation for titanium metal [21, 22] and Ti-based alloy Ti-Zr-Cu-Pd [23] without any suggestions about the origin of their appearance (see **Figure 6**). From our opinion the production mechanism of quasi-grating **G** superimposed on the $\mathbf{g} \parallel \mathbf{E}$ resonant grating is analogous for one suggested for nanosecond irradiation regime. Note that the thermal etching groove formation is inherent to (poly)crystalline materials, but the alloy Ti-Zr-Cu-Pd is the amorphous one. In fact under the alloy heating up to high temperature, the alloy transfers from its metastable state to the crystalline one, and the suggested model works further.

http://dx.doi.org/10.5772/66248

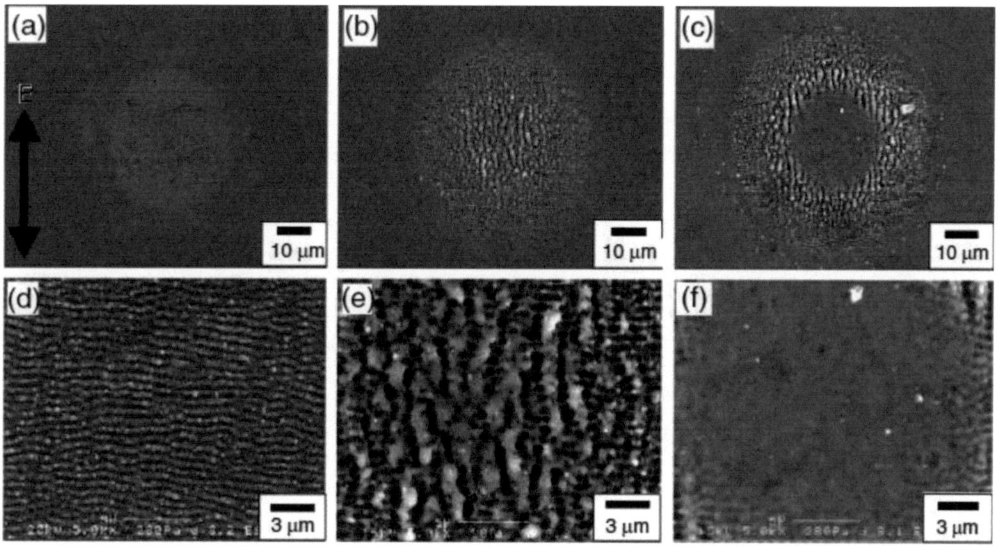

Figure 6. The quasi-periodic microrelief obtained on a titanium surface under the action of ultrashort pulses (τ = 150 fs, λ = 800 nm, pulse-repetition rate 1 kHz, N =10) for energy densities Q = 0.25 J/cm^2 (a) and (d), 0.75 J/cm^2 (b) and (e), and 1.5 J/cm^2 (c) and (f) at low (upper row) and high magnification (lower row), $\mathbf{g} \perp \mathbf{E}_t$ [21].

In one of the previous models proposed for explanation of spatial periodic relief on metal surfaces under ultrashort laser illumination [24], the process was considered with nucleation in a distended metal and the ensemble of cavitating bubble self-ordering [24]. According to authors [24] and physics of the process, the orientation and periods of produced nanostructures are independent on laser radiation polarization and wavelength, accordingly. The typical power densities of laser radiation must be higher than in our consideration. In this model, the predicted spatial periods are of the laser heating depth (10 ÷ 100) nm order. Hence, the model [24] is not applicable for our case.

3. Effect of electroplasticity

The effect of electroplasticity will be considered in this section as one of the most closed volume analog for anisotropic grain boundary movement (AGM) effect (Section 2). At first the electroplasticity effect (EPE) in metals was discovered by Troitsky [25]. The metals demonstrate the enhanced plasticity under the influence of the high directed current density.

The high perfection of axial texture is formed due to metal grains of definite crystallographic orientations turned in the direction of wire dragging (**Figure 7**). The initial material takes the texture "suitable" for subsequent high current transfer [26]. This structure turns out more perfect than that formed by the usual wire-dragging technology. During electro-plastic wire dragging, the deformation strengthening fails and the plasticity increases as a result [27–35].

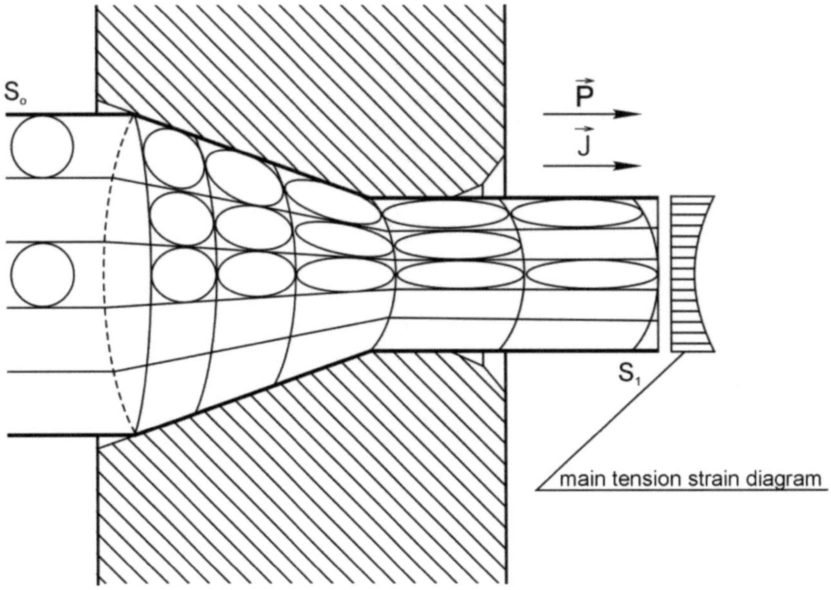

Figure 7. The perfect wire texture creation during electro-plastic wire dragging in condition of current density **j** and wire movement direction **P** vector coexistence [26].

The thorough analysis shows that the nonthermal current action mechanisms involve electro-plastic effect and ponderomotive forces of pulsed current and impact the wire tension decrease sufficiently. More efficient pulsed current action (in comparison with dc) [26, 30], current polarity dependence evidence the nonthermal nature of the occurred processes. In addition, during rapid wire dragging (>10 m/s), the Stewart-Tolman effect takes place (the delay of free electron gas against crystal lattice accelerating in the area of metal deformation inside a drawing ring caused by transition to a smaller diameter). The Stewart-Tolman effect is electronic by nature, and it favors the axial wire texturing to the electro-plastic and its degree of perfection increasing. The typical parameters realized under the electro-plastic effect are listed below:

- Frequency of current flow (10 kHz)
- Current power density (250 kA/cm^2)
- Current pulse duration (60 μs)
- Wire-dragging velocity (rapid dragging) (≥10 m/s)

The observed effect is caused by plasticity influence of EPE in metal volume; it relieves the process of axial texture formation and decreases friction in dragging die due to the grain crush in the near-surface wire areas.

Apart from the perfect grain structure creation for copper, it was observed as follows [26]:

- The number of randomly distributed dislocations decreases.

- Wire electrical resistivity decreases by 8–15%.
- Wire residual plasticity increases by 25–30%.
- Wire-dragging force decreases by 25–30%.

For the stainless steel wire, it was observed as follows [26]:

- Electrical resistance decreases by 18–20% in comparison with warm wire dragging and by 26–30% in comparison with cold wire dragging without current.
- Residual plasticity increases by 5–20%.
- The strength limit decreases by 15–20%.

The electroplasticity effect is a linear function of current density and was never observed for alternated current. It is known that to remove inner tension and residual deformations on high-voltage transmission equipment, the powerful pulse current is transmitted by the wires hanged preliminary before the wires are to be arranged finally.

So, in general, the electroplasticity effect may be considered as a volume analog of the effect of anisotropic grain boundary movement (AGM) being discussed. Really, the electroplasticity effect is caused by directed current of electrons' action. And the AGM effect is also the consequence of the direct current of electrons forced by the effect of electrons drag by surface plasmon polaritons in optical skin layer of metal. In both cases, the directed electrons transfer their momentums to the grain boundary as a wall and force their displacement.

4. Conclusion

The formation of microstructures on metal surface under the interaction of laser radiation of nanosecond durations in regime of near-threshold melting was theoretically and experimentally analyzed. The generations of regular structures in mode of linear gratings or normal and abnormal orientations the periods of which are proportional to laser radiation wavelength were observed. The mechanisms of discussed structure formation are in frameworks of universal polariton model of laser-induced condensed mater damage. In the regime of power density not exceeding the melting threshold, the relief of a new type in mode of quasi-grating of grooves with periods of the order 5 μm and grating orientation orthogonal to the electric field strength vector of laser radiation has been studied. So, the anisotropy of grain growth process was observed. To explain the result of anisotropic growth, the qualitative model was suggested. The model is based on the effect of skin-layer electrons dragged by surface plasmon polaritons. The flux of electrons transfers its momentum to the grain boundary as to the wall given an anisotropic force. So, the produced skin-depth surface layer has properties differing from the bulk of metal.

The considered effect is the surface one. As the volume analog to this effect the well-known electroplasticity, one can be considered. Really, in both cases, the cause of the main effect in metal is the directed flux of electrons.

Author details

Makin Vladimir Sergeevich

Address all correspondence to: makin@sbor.net

Scientific Research Institute for Opto-Electronic Device Instrumentation, Sosnovy Bor City, Leningrad Oblast, Russia

References

[1] Bonch-Bruevich A.M., Libenson M.N., Makin V.S., Trubaev V.V. Surface electromagnetic waves in optics. Opt. Eng. V. 31. No. 4. Pp. 718–730 (1992).

[2] Makin V.S., Makin R.S. Nonlinear interaction of linear polarized laser radiation with condensed media and overcoming the diffraction limit. Opt. Spectrosc. V. 115. No. 2. Pp. 162–167 (2012).

[3] Makin V.S., Pestov Y.I., Makin R.S. Abnormal spatial nanogratings formation by long pulse laser radiation on condensed matter surfaces. Proceedings of International Conference "Days on Diffraction". 2016. Pp.298–303.

[4] Makin V.S., Logacheva E.I., Makin R.S. Origin of anomalous nanostructures formation under linear polarized femtosecond laser irradiation of condensed matter. Proceedings of International Conference "Days on Diffraction". 2015. Pp. 201–207.

[5] Makin V.S., Logacheva E.I., Makin R.S. Localized surface plasmon polaritons and nonlinear overcoming of the diffraction optical limit. Opt. Spectrosc. V. 120. No. 4. Pp. 610–614 (2016).

[6] Makin V.S., Makin R.S. Interaction of laser radiation with an axially symmetric polarization with condensed media. Opt. Spectrosc. V. 115. No. 4. Pp. 591–595 (2013).

[7] Olemskoĭ A.I. Synergetic of complex systems. Moscow. Kasandr. 2009.

[8] Schmidt S., Nielson S.F., Gundlach C., Margulies L., Huang X., Juul J.D. Watching the growth of bulk grains during recrystallization of deformed metals. Science. V. 305. P. 229 (2004).

[9] Han-Riege C.S., Thompson C.V. Microstructural evolution induced by scanned laser annealing in Al interconnects. Appl. Phys. Lett. V. 75. P. 1464 (1999).

[10] Gulyaev A.P. Physical metallurgy. Moscow. Mashinostroenie. 1986.

[11] Makin V.S., Makin R.S., Vorobiev A.Y., Guo C. Feigenbaum's universality and Sharkovsky order in laser-induced periodical structures at interfaces and in bulk of

condensed matter. In Nonlinearity in modern nature. Editor Malinetskii G.G. Moscow. LKI Publisher. 2009. Pp. 303–322 (in Russian).

[12] Libenson M.N., Bonch-Bruevich A.M., Makin V.S. Surface polaritons and powerful radiation action. Usp. Fiz. Nauk. V. 155. P. 719 (1988) [Sov. Phys. Usp. 31, 772 (1988)].

[13] Samblse J.R. Grain-boundary scattering and surface-plasmon attenuation in noble-metal films. Solid State Commun. V. 49. P. 343 (1984).

[14] Johnson P.B., Christy R.W. Optical constants of the transition metals: Ti, V, Cr, Mn, Fe, Ni, and Pd. Phys. Rev. B. V. 9. P. 5056 (1974).

[15] Mikheev G.M., Nasibulin A.G. Photon drag effect in single-walled carbon nanotube films. Nano Lett. V. 12. No. 1. Pp. 77–83 (2012).

[16] Durach M., Rusina A., Stockman M.I. Giant surface-plasmon-induced drag effect in metal nanowires. Phys. Rev. Lett. V. 103. P. 186801 (2009).

[17] Geguzin Y.E., Ovcharenko N.N. Surface energy and surface processes in solids. Usp. Fiz. Nauk. V. 76. P. 283 (1962) [Sov. Phys. Usp. 5, 129 (1962)].

[18] Bozhevolnyi S.I., Volkov I.S., Devaux E., Laluet J.-K., Ebbesen T.W. Channel plasmon polariton guided by subwavelength metal grooves. Phys. Rev. Lett. V. 95. P. 046802 (2005).

[19] Chiappe D., Toma A., de Mongeo F.B. Tailoring resistivity anisotropy of nanorippled metal films: electron surfing on gold waves. Phys. Rev. B. V. 86. P. 045414 (2012).

[20] Makin V.S., Makin R.S. Lateral relativistic electron beam synergetic creation and transport by petawatt laser radiation. Proceedings of International Conference Days on Diffraction, May 30–June 3, 2011, St-Petersburg, Russia. Pp. 133–136.

[21] Tsukamoto M., Asuka K., Nakano H., Hashida M., Katto M., Abe N., Fujita M. Periodic microstructures produced by femtosecond laser irradiation on titanium plate. Vacuum. V. 80. P. 1346 (2006).

[22] Vorobyev A.Y., Guo C.L. Femtosecond laser structuring of titanium implants. Appl. Surf. Sci. V. 253. P. 7272 (2007).

[23] Shinonaga T., Tsukamoto M., Mariyama S., Matsushita N., Wada T., Wang X., Honda H., Fujita M., Abe N. Femtosecond and nanosecond laser irradiation for microstructure formation on bulk metallic glass. Trans. JWRI. V. 38. No. 1. P. 81 (2009).

[24] Zhakhovskiĭ V.V., Inogamov N.A., Nishihara K. New mechanism of the formation of the nanorelief on a surface irradiated by a femtosecond laser pulse. Pis'ma Zh. Eksp. Teor. Fiz. V. 87. P. 491 (2008) [JETP Lett. 87, 423 (2008)].

[25] Troitsky O.A. Electromechanical effect in metals. Pis'ma v JETF. V. 10. P. 18 (1969).

[26] Troitsky O.A., Stashenko V.I., Ryszkov V.G., Lyashenko V.P., Kobyl'skaya E.B. Electroplastic wire-dragging and new technologies to simplify wire creation. Problems of

Atomic Science and Techniques. Seria: Physics of Radiative Damage and Radiative Material Science. V. 4. No. 98. Pp. 111–117 (2011).

[27] Gromov V.E., Zuev L.B., Kozlov E.V., Tsezermayer V.Y. Electrostimulated plasticity of metals and alloys. Moscow. Nedra. 1996. 280 p.

[28] Roschupkin A.N., Troitsky O.A., Spitsin V.I., et al. The high density current action on metal plastic deformation conception development. Dokl. AN SSSR. V. 286. No. 3. Pp. 633–636 (1986).

[29] Mel'nikova N.V., Hon Y.A. To the theory of electroplastic deformation of metals. Phys. Mesomech. V. 3. No. 5. Pp. 59–64 (2000).

[30] Dubinko V.I., Kras V.I., Klepikov V.F., Ostapchuk P.N., Potapenko I.F. Enhanced Plasticity under the electric current pulses action modeling. Problems of Atomic Science and Technique. Series: Physics of Radiative Damage and Radiative Material Science. V. 2. No.4 Pp.158–166 (2009).

[31] Kravchenko V.V. Directed electron's flux action on moving dislocations. JETF. V. 51. Pp. 1676–1681 (1966).

[32] Malygin G.A. Self-organization dislocations process and plasticity of crystals. Usp. Fiz. Nauk. V. 169. No. 9. Pp. 979–1008 (1999).

[33] Spitsin V.I., Troitsky O.A., Ryzhov V.G., Kozyrev A.S. On-draw-plate copper wires dragging. Dokl. AN SSSR. V. 231. No. 2. Pp. 402–404 (1976).

[34] Spitsin V.I., Troitsky O.A. Electro-plastic deformation of metals. Moscow. Nauka. 1985. 160 p.

[35] Troitsky O.A. Ultrasound electroplastic flatting of metals. Bulletin for Science-Technology Development. National Technology Group. V. 10. No. 26. Pp. 42–49 (2009).